MW00390585

Wild
Roots

Wild Roots

A Forager's Guide to the Edible and Medicinal Roots, Tubers, Corms, and Rhizomes of North America

DOUG ELLIOTT

Healing Arts Press
Rochester, Vermont

WARM, FRIENDLY, ACCOMODATING PEOPLE

The following is a partial list of people I would like to thank:

Nancy, Joe, Grant, Carol, Laura, Fred, Kay, Connie, Marina, Henry, Sugar Britches, Natalie, Greg, Carol, Don, Ellen, Norm, Wendy, Chris, Jean, Brynda, Dave, Carol, Frank, Downey, Don, Fuller, Doug, Walt, Bottle Hill, Sassafras Kitchen, Larry, Veda, Raymond, Linda, Pat, Jill, and Pigweed.

I've had help, encouragement, and companionship from Lynn, Helen, and Sweet Cicely that is beyond thanks.

Don Mcleod, Jerry Jenkins, and Dan Pittillo allow me to barrage them with botanical questions.

Jerry Jenkins, Ellen and Don Elmes, and Bruce Williams offered a great deal of editorial advice.

Mother did the typing, for which I am most grateful.

Healing Arts Press
One Park Street
Rochester, Vermont 05767

Note to the reader: This book is intended as an informational guide. The remedies, approaches, and techniques described herein are meant to supplement, and not to be a substitute for, professional medical care or treatment. They should not be used to treat a serious ailment without prior consultation with a qualified healthcare professional.

LIBRARY OF CONGRESS CATALOGING-IN-PUBLICATION DATA
Elliott, Douglas B.
Wild roots : a forager's guide to the edible and medicinal roots, tubers, corms, and rhizomes of North America / Doug Elliott.
p. cm.
Originally published: Roots. Old Greenwich, Conn. : Chatham Press, c1976.
Includes bibliographical references and index.
ISBN 0-89281-538-8
1. Medicinal plants—North America—Identification. 2. Wild plants, Edible—North America—Identification. 3. Roots (Botany)—Identification. 4. Plants—Folklore.
I. Elliott, Douglas B. Roots. II. Title.
QK99.N67E445 1995
581.6'3'097—dc20 95-13383
CIP

Printed and bound in the United States

10 9 8 7 6 5 4 3 2 1

Healing Arts Press is a division of Inner Traditions International

Distributed to the book trade in Canada by Publishers Group West (PGW), Toronto, Ontario

Distributed to the health food trade in Canada by Alive Books, Toronto and Vancouver

Distributed to the book trade in the United Kingdom by Deep Books, London

Distributed to the book trade in Australia by Millennium Books, Newtown, N.S.W.

Distributed to the book trade in New Zealand by Tandem Press, Auckland

CONTENTS

The Beginning of The Book

The first summer devoted to full-time gathering and studying of herbs was spent in northern New England, at the foot of the White Mountains in New Hampshire and along the rockbound coast of Maine. I was like the proverbial grasshopper fiddling away those brief, blissful, northern summer days . . . sampling the bountiful herbs and flowers of the bogs, mountains, marshes and meadows.

When autumn approached, and the mountain of undone winter preparations loomed over me, the grasshopper mentality persisted. Yet, a certain extremity provided me an alternative that not one of the six legs of the grasshopper provided him—I had a thumb. With my trowel, plant books and much neglected sketch pad in my pack—and thumb extended—I headed down the Appalachian Mountain chain to West Virginia, to be welcomed into the ridge-top home of my friends, Don and Ellen. They too were enjoying the "easy" country life, shoveling away at their own mountain of winter preparation chores. They didn't mind catering to some unseasonal fiddling though and introduced me to Joe, their 73 year old neighbor from the next hollow, who took me behind his house to show me ginseng.

In subsequent wanderings through this area more ginseng was found, as well as other classic Appalachian medicinal roots. The lore behind these roots is entrancing; their forms and structures are intriguing. I dug some roots and washed them with particular care, then got out the sketchbook and black ballpoint pen. Almost before I realized it, there on the paper appeared some of the most sensitive renderings I had ever done, hardly having touched a pen to draw for almost a year. The magic of this subterranean scenery had taken hold. Now, I want to share this experience with you.

Plant Names

We don't have to learn the name of a plant to appreciate it. A rose by any other name is still a rose. But if we *do* appreciate a plant we may want to talk about it, or look it up and read what others have observed about it. For this we need to know the name.

Therefore, I have listed most of the common names a plant has been given, as well as its current scientific label — to avoid confusion. For example, there are at least a dozen different plants in the East called "snakeroot." If you wanted to discuss the plant with someone, without having this snakeroot in front of you both, unless you had the scientific name you could not be sure that you were both talking about the same species.

The "latin name" is interesting on another level. Not only does the scientific nomenclature specifically label the plant, but it also states a relationship between that plant and the rest of the plant kingdom. The divisions of the scientific name are: family, genus, species. These indicate plant characteristics, not importance or parentage. When botanists first gave names to a family of similar plants, they usually used the name of one of the family members, e.g., the family *ROSACEAE* is named for the genus *ROSA*. Because of this naming policy, the family comes to be called the "rose family" as if the rose is the "head" of the family, even though strawberries, blackberries, apples and cherries are equally important members of *ROSACEAE*. It might be helpful to remember that when someone speaks of a plant being in the Arum family, for example, they mean it is in the same family as arum, which happens to be called *ARACEAE*.

The major subdivisions within a family are the genera. This generic name is like a last name or surname, which differs among each family in the "clan." The further subdivision into species describes a particular individual, with a particular characteristic. For example, in the name *Rosa multiflora: Rosa* is the genus, *multiflora* indicates the species and means "many flowered."

Dilution and Evolution

As early as 1420 Fra Tomás de Berlanga proposed that vegetation is simply animal life turned inside out. The leaves conduct respiration and correspond to lungs. The roots are the unencased digestive tract which takes in and assimilates nutrients. Sunlight is the energizing heart, he concluded, without which the whole system would fail to operate.(1)

More recently, a similar comparison has been made by Emanual Epstein of the University of California.(2) He points out that the evolutionary tendencies in a higher animal, such as a rabbit or a human, tend to minimize their surface area as much as possible. This keeps the organism compact, which makes body temperature regulation more efficient and also exposes the least possible surface area to injury. With a plant the opposite is true. Plants consist of extensive systems that expose massive amounts of surface area to the environment, both above and below ground (except in certain cacti and desert plants, where surface area also means an evaporation of valuable moisture).

The maintainance of such large surface areas is necessary because the nutrients and energy sources that plants use are very diffuse. The leaves must absorb not only carbon dioxide, which is found only in very low concentrations in the atmosphere, but also diffuse radiant energy from the sun. The more surface area a plant maintains, the more of these two dilute resources it will be able to intercept.

The same principle of large-surface-area is in effect with the roots, but even more so. One of the primary functions of a root is to absorb nutrients from the earth, in the form of very dilute mineral salts in the soil's water. Since the soil and its solutions are essentially unstirred, the root must absorb substances that are relatively immobile. Whereas the leaf surface, in contrast, is rarely in prolonged contact with the same air molecules.

The evolutionary response to the dilution and immobility of soil nutrients has produced root systems that branch and rebranch in order to bring an incredibly large surface area in contact with the soil. At the smallest rootlets, the surface area is further multiplied by a covering of root hairs. Root hairs are absorbtive epidermal cells that extend out laterally from the root.

Like an iceberg, there is much more beneath the surface than you would believe by looking at what extends above the surface. A classic experiment illustrating the dimensions attainable by root systems was conducted in the 1930's. Howard J. Dittmer at the State University of Iowa planted a winter rye plant (Secale cereale) in a box of soil 12 inches square and 22 inches deep. After four months of growth the plant was 20 inches high and consisted of a clump of eighty shoots. Dittmer carefully liberated the root system from the soil and measured it. Adding the lengths of the roots together, the total length, not including the root hairs, was 387 miles and the surface area was 2,554 square feet. When the length and surfaces of the root hairs themselves was also estimated and added, the total was nearly 7,000 miles in length and 6,500 square feet in surface area for the entire root. As a comparison, the surface area of the above ground part of the rye plant amounted to an area of 51.4 square feet.(3) Many plants have very deep as well as multiple roots. The prairie plants in particular, because of their need to withstand exposure to prolonged dry seasons, descend to surprising depths. For example, the roots of buffalo berry (*Sheperdia argophylla*), dotted button-snakeroot (*Liatris punctata*), and alfalfa (*Medicago sativa*), have all been known to reach depths of 20 to 50 feet. It is also known that the roots of some long-lived forest trees descend to depths of 100 feet.

To maintain the intimate relationship between the roots and the soil, the plant, of course, must remain stationary. A stationary organism has no need for sophisticated sense organs, complex body motions, the ability to make decisions, or a complex central nervous system. The plant kingdom, through the course of evolution, has had no need for what scientists call "intelligence." Interestingly, from the point of view of a plant (if a plant could have a point of view) "intelligence" might appear as: the crutch used by an organism which is not able to absorb its own nutrition directly from its environment. The plant possesses a certain wonderful sophistication here.

Actually, it seems that some humans have also mastered this ability to sustain life by directly absorbing energy. There are several, well documented cases of humans who lived long, healthy lives without eating gross food of any kind. Two of these individuals lived during the twentieth century. One was yogi Giri Bala of Bengal, India, who through the use of yogic breathing exercises and techniques lived for more than fifty years without food, deriving nourishment directly from the energies of air and sunlight.(4) The other was the Catholic mystic, Therese Newmann of Konnersreuth, Germany, who lived nearly forty years without eating. As she put it, she lived solely "by God's light."(5)

Perhaps these individuals represent the vanguard of human evolution; gaining freedom from food gathering and all that entails, without giving up their freedom of mobility and intelligence.

Scenery Beneath the Greenery

The underground organs of plants are divided into several categories. First, there are *true roots*, which play several different vital roles in the life of the plant. They anchor the plant in the soil. They absorb water and minerals from the soil. They conduct these nutrients to the stem, which in turn transports them to the rest of the plant. They can also serve as receptacles for food storage.

The true roots are generally thought of as two separate categories: 1. Those that are derived from the primary, i.e. the seminal or seed root. 2. Those that are derived in some other way, called adventitious roots.

As the seed germinates, the *primary root* is the first root that comes out of the seed. Such primary root systems that have a central root from which the others radiate are called *tap roots*. Tap roots are often fleshy, thickened, and serve to store food, usually in the form of starch (Fig. 1). On the other hand, primary roots that divide at once into a cluster are called *fibrous roots* (Fig. 2). These can also be swollen, fleshy and tuberous as in the case of dahlia, or day lily (see page 78).

Adventitious roots are derived from other than the primary seed root. A good example is the prop roots of corn, or the roots that form when the stem of a plant such as blackberry touches the ground and reroots.

The second large group of underground plant organs is the *rootstocks*. In essence they are underground stems that produce both leaves and true roots. However, they serve as organs of storage, absorption, conduction and anchorage—so they have the same function as roots. In herbal nomenclature the various forms of rootstocks are referred to simply as roots.

These underground stems take on a variety of forms: The *rhizome* (Fig. 3) is a horizontal, creeping stem usually producing roots from the nodes and buds in the leaf axils. The leaves are often present as non-chlorophyllous scales. *Tubers* (Fig. 4) are the swollen tips of stems, while a *corm* (Fig. 5) is the enlarged base of a stem. A *bulb* (Fig. 6) is a storage organ made up of scales which are actually thickened leaves.

Now that I have enumerated all of these neat, orderly categories, it is comforting to know that a group of organisms as diverse and variable as the higher plants cannot, in fact, be so easily confined to these concepts.

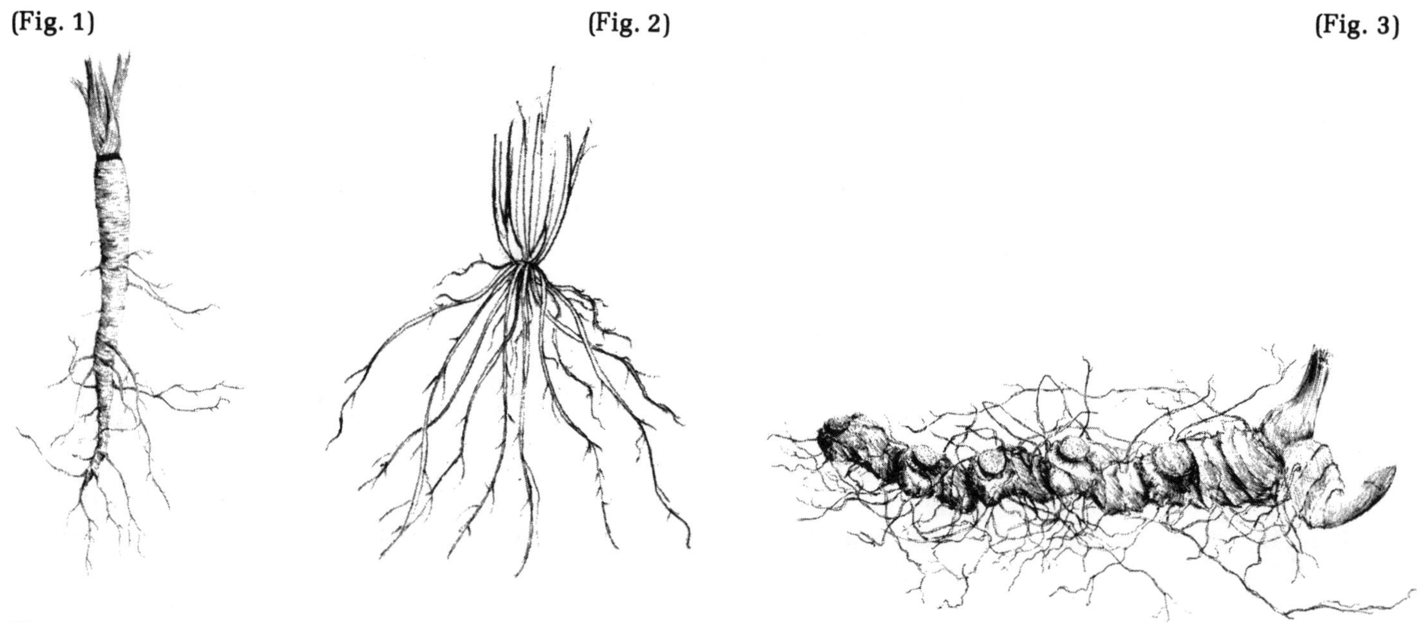

(Fig. 1) (Fig. 2) (Fig. 3)

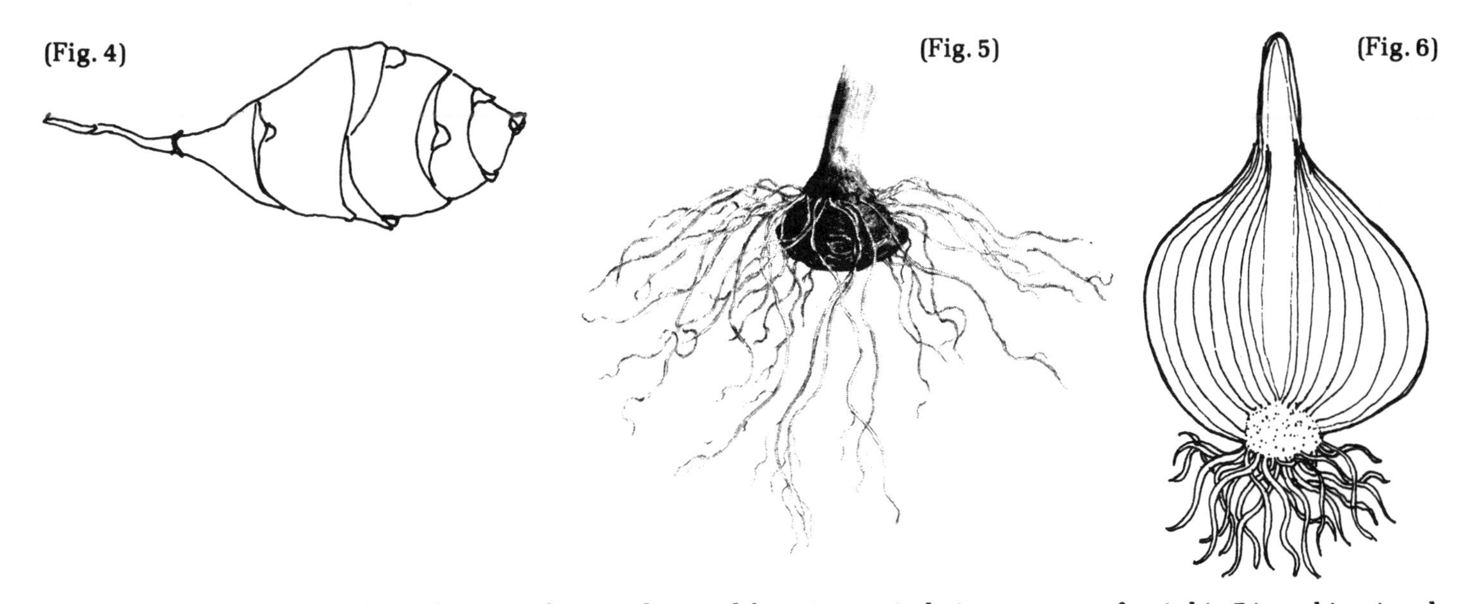

Roots are not necessarily underground. Some live and function entirely in water: e.g. frog's bit *(Limnobium)* and duckweed *(Lemna)*. Some roots, like those of the *Angreaecum* orchids, grow in air and do anchor the plant, but they function more like leaves. The sticky mistletoe *(Phoradendron)* seed adheres to the bark of a host tree and as it germinates its root squeezes right through the bark into the sap stream of the host.

The dodder or love vine *(Cuscuta)* germinates from a seed on the ground. Rooted at first in the soil, it develops a slender leafless stem. By twisting about, the stem eventually comes into contact with the right host where upon it latches on and establishes intimate contact with the sap stream by means of adventitious roots called *haustoria*. (Hence the name "love vine.") As soon as the haustoria are established, the growth of the dodder's primary root system stops and eventually withers away.

And then there are plants like spanish moss *(Tillandsia usneoides)* which is neither moss nor spanish but kin to the pineapple, that have no roots at all and can grow just as easily hanging from a telegraph wire as from a tree branch.

Machinery Beneath the Greenery

The growing root is a hotbed of activity. It has been recorded that the root of corn, for example, grows at a rate of more than two inches per day for three or four weeks. The root tip secretes a weak acid which chemically breaks down some of the surrounding soil particles. By secreting this acid the root not only releases minerals from the soil for the plant to later absorb, but it also "eats" a path through the soil.

The root tip is covered with a thimble-like shield of cells called the root cap, which protects it from injury as the root pushes forward through the soil particles. The cells of the cap are continually worn away and sloughed off. This results in a slimy covering on the adjacent soil particles, again facilitating the passage of the root through the soil.

The sloughed off cells, along with other exudations of the root, attract and nurture other organisms, such as bacteria. It has been found that a healthy root often has colonies of bacteria which form a sheath-like cover, several bacteria deep, over the entire root. The bacteria populations seem to stimulate the root's growth and nutrient uptake. "Commensal" is the word that biologists use to describe such a mutually beneficial relationship as that which roots maintain with other tenants of the soil.

Just behind the growing tip of the root appear long tubelike epidermal cells called *root hairs*. These root hairs act to increase the soil contact surface area of the root, thus increasing its absorptive powers. It has been estimated that on the hair-bearing portions of the roots of a common pea about 1400 hairs occur on every hundredth of a square inch of surface. However, these root hairs only develop near the tip of the root. As the root works its way through the soil, new root hairs are formed at the younger end of the root. The older hairs farther back on the root collapse and slough away. Some root hairs live only a few hours and others last more than three or four days.

As a replacement for the dead root hairs, most roots are invaded by fungi called mycorrhizae ("fungus-roots"). Mycorrhizal associations are mutually beneficial, or commensal, as long as a balanced relationship is maintained. Nutrients are absorbed from the soil by the fungus and released to the host cells. The fungus in turn obtains food from the host plant. The degree of benefit that each partner receives varies. Plants that lack chlorophyll may be completely dependent on the fungi to supply them with both organic and inorganic nutrients. While other plants in fertile soil are able to absorb nutrients and achieve optimum growth without the help of fungi.

There are basically two kinds of mycorrhizae. Both kinds have hyphae, i.e. fungus branches or filaments that penetrate into the root. The hyphae of one kind grow between the cells of the root. This kind is called ectomycorrhizae. The other kind, called the endomycorrhizae, have hyphae that actually grow into the cells themselves.

We'll take the *ectomycorrhizae* first. This type of fungus primarily infects the roots of trees and shrubs, particularly birch, pine and beech. The fungus forms a compact sheath over the roots, and the hyphae from the sheath penetrate between both the epidermal and cortical cells to form an interlocking network called the Hartig net. Also, the hyphae strands grow out into the soil and act as root hair-like absorbing surfaces. Once nutrients are absorbed they are conducted into the root and passed on to the plant cells through the Hartig net. These ectomycorrhizal roots are swollen due to fungus-produced growth hormones (auxins).

There are many species of ectomycorrhizal fungi. Some will only infect certain species of plants, while others may be found on a wide variety of hosts. Most ectomycorrhizal fungi are deficient in certain enzymes and are unable to break down and utilize lignin and cellulose. They obtain these from the host plant and it is believed that these fungi are incapable of existing apart from living roots. The plant host, in turn, is benefited by the fungus infections in a number of ways. An ectomycorrhizal root is thicker, it is stimulated to branch more, it lives longer than its nonmycorrhizal counterpart, and it has external hyphae that function as root hairs. Each of these characteristics increase the root's absorptive surface, and its potential to take in nutrients. Also, the ectomycorrhizae remain in one place for a relatively long period of time while the uninfected root tip moves, i.e. grows on. Because of the time necessary to establish an ion flow (a flow of nutritive mineral salts through the soil) an organ that remains active in the same position longer can obtain a relatively larger volume of nutrients. The fungal sheath forms a barrier to disease-causing organisms, both mechanically and by antibiotic secretions. Only under special highly fertile conditions can a plant that is normally ectomycorrhizal be grown without mycorrhizal fungi and attain normal growth. For instance, it is usually difficult to grow trees in a new area where their native mycorrhizae are not present.

The *endomycorrhizae* are variable as to their shapes and roles in plant growth. Many of them form hyphal sheaths like the ectomycorrhizal fungus but their hyphae penetrate into (rather than between) the cells of the cortex and form either inner coils or branched tree-like structures that are digested by the host cell. In this manner they exchange nutrients with the host plant.

These fungi have remarkable associations with certain plants. The genus *Monotropa*, of which the ghostly Indian pipe is a member, lacks chlorophyll and depends entirely on a mycorrhizal fungus for its organic compounds. This fungus forms ectomycorrhizal association with nearby tree roots, links up endomycorrhizally with the roots of *Monotropa*, and transmits nutrients from the tree to the *Monotropa*. This makes *Monotropa* a parasite by proxy, sometimes called an epiparasite. Parasitic or not, it seems that each party benefits by the association! the trees in the ways mentioned previously, the *Monotropa* by receiving its essential nutrients, and the fungus, it is believed, by obtaining a growth promoting factor from the *Monotropa*. To each, his due.

Orchids are also highly dependent on mycorrhizal infection for normal growth and survival. Orchid seeds are so tiny and fragile that under natural conditions they cannot germinate and grow unless the embryo becomes infected. As adults, the terrestrial species with large, chlorophyll-containing green leaves and well developed roots, like lady slipper (Cypripedium), may not maintain mycorrhizal associations. However, those orchids that lack chlorophyll *(corallorhiza)* remain dependent upon mycorrhizal associations throughout their life.

Some orchid mycorrhizae cause disease in other plants. In this way the fungus can protect its host from competition. The same fungus organism that is disease-causing on a tree may at the same time form mycorrhizae with nearby orchids. In this orchid-fungus relationship too much or too little of a good thing can be fatal. A delicate balance must be maintained at all times or the fungus can become pathogenic, overcoming and eventually killing its host. But the orchid has two ways of controlling the fungal growth. One is by digesting the hyphal coils of the fungus, and in the process obtaining nutrients; or by producing certain substances in a tuber-like collection that are toxic to the fungus. This checks the fungus if it starts to spread.

Endomycorrhizal fungi are of economic importance when they infect crop plants. In Illinois, soybean roots are highly mycorrhizal even on plants grown in fertile soil. Changes in the mycorrhizal infection effect crop yield—a crucial fact to the farmer. Sometimes after fumigation of a field the growth of the crop is retarded. This has been attributed to "soil toxicity," although further investigation indicates that this could be due to the killing off of mycorrhizal fungi.

Aquatic plants and other plants growing in very wet environments are usually reported to be without mycorrhizae. Rice, when growing in a wet habitat, is non-mycorrhizal; but when it is planted in a drier soil it forms mycorrhizae. There do seem to be relatively few aquatic commensal fungi, perhaps because of the oxygen deficient silt that is the habitat of most pond and marsh plants.

Affix
Postage
Stamp
Here

Inner Traditions International, Ltd.
P.O. Box 388
Rochester, VT 05767
U.S.A.

If you wish to receive a copy of the latest INNER TRADITIONS INTERNATIONAL catalog and to be placed on our mailing list, please send us this card. It is important to print your name and address clearly.

Date ____________________

Name ____

Address ____

City ______________________________________ State ________ Zip ________

Country ____

I

Roots of Shade-Loving Forest Plants

The roots of deep forest plants anchor the understory growth beneath the tall canopy of shade trees on the hills and in the hollows of our eastern woodlands. Plants that thrive in such light-starved environments have evolved certain general characteristics. Most of them have thickened perennial rhizomes that, during the growing season, store starch and form the bud or buds of the next year's growth. As soon as the first warmth of spring arrives, the plant takes advantage of this reservoir of stored energy and "bolts" out of the ground to get on with its flowering and related vernal activities well before the trees leaf out, intercepting the life giving sunlight. This is why a woodland wild flower walk is more rewarding in early spring than at any other time of the year. Of course, there are notable exceptions, like black cohosh which blossoms in mid-summer well after the canopy has filled out. One of the adaptations that assures black cohosh's survival is its relatively large size and its particularly wide spreading leaves that give it a much better chance of intercepting the little sunlight that does filter through.

Geranium maculatum

Geraniaceae, **Geranium family**

OTHER COMMON NAMES:
Alum Root, Crane's-bill, Crowfoot, Stork's-bill, Dove's-foot, Old Maid's Night-cap, Shameface

WILD GERANIUM

Wild Geranium is found in moist, wooded areas, from Eastern Canada south to Missouri and Georgia.

This attractive woodland plant attains a height of two feet, with soft, hairy leaves palmately dissected into three- to five-toothed lobes. It is the shape of the leaves that give the plant the name crowfoot. The name dove's-foot probably originated in England where a plant of a different species than our Wild Geranium, but with similar appearance and properties is also called dove's-foot. The delicately veined, five-petaled, blushing pink to rose-purple flowers, which bloom from April to June, give the plant its name of shameface. The shape of the slightly wilted blossom, when held upside down, somewhat resembles a lady's old-fashioned sleeping bonnet and probably accounts for the name old maid's night-cap.

As spring moves into summer, the flower forms into a five-sectioned, beaked seed capsule from which the names crane's-bill and stork's-bill is derived.

Wild Geranium is most known and valued for its astringent properties. Now, whenever someone doesn't understand the concept of astringency in herbal medicine, I have him chew a piece of Wild Geranium, and in minutes he perfectly understands what would have taken much effort to explain. Clearly a case of one "chaw" being worth a thousand words.

The word "astringent" comes from the Latin *astringere*, "to bind fast," and refers to the ability of a substance to draw together or tighten soft organic tissue. Alum, an aluminum sulfate compound, has been used for its astringency. Wild Geranium, with its high tannin content, got the name alum root from astringent properties similar to alum. However, the plant contains no appreciable amount of aluminum or sulfur, and cannot be substituted for the chemical alum used as a mordant in vegetable dyeing.

Because of Wild Geranium's powerful, non-irritating, astringent properties, it is a classic treatment for all types of weak, atonic conditions of the organs or tissues, especially those accompanied by excessive discharges. This would apply to diarrhea, dysentery, and vaginal discharge. It is also considered a good wash or gargle for canker sores, inflamed mouth or gums, and sore throat. Because of its ability to tighten and impart tone to relaxed and enfeebled tissues, it is considered a tonic.

Wild Geranium is recommended in combination with a small amount of cayenne pepper to tone and reactivate the weakened stomach of hard drinkers. It is also used as a styptic, i.e. an astringent aimed at stopping bleeding, for all kinds of hemorrhages from simple nose bleed to piles, or for more serious internal bleeding. It is administered in the form of a tincture or an infusion of the root.

The rootstock is up to six inches long, bearing numerous knobs out of which smaller rootlets grow. The fact that Wild Geranium is a perennial is evident from the stem scars showing previous years' growth, and the bud forming at the base of the present year's stem. When fresh, the root is light brown externally and inside is light-colored and somewhat fleshy. When dried, it turns hard and wrinkled with a grey-purple color inside. It has been found that the tannin content in Wild Geranium is highest just before the plant flowers, so it is best gathered for herbal use in the spring rather than the fall, as is the case with most other roots.

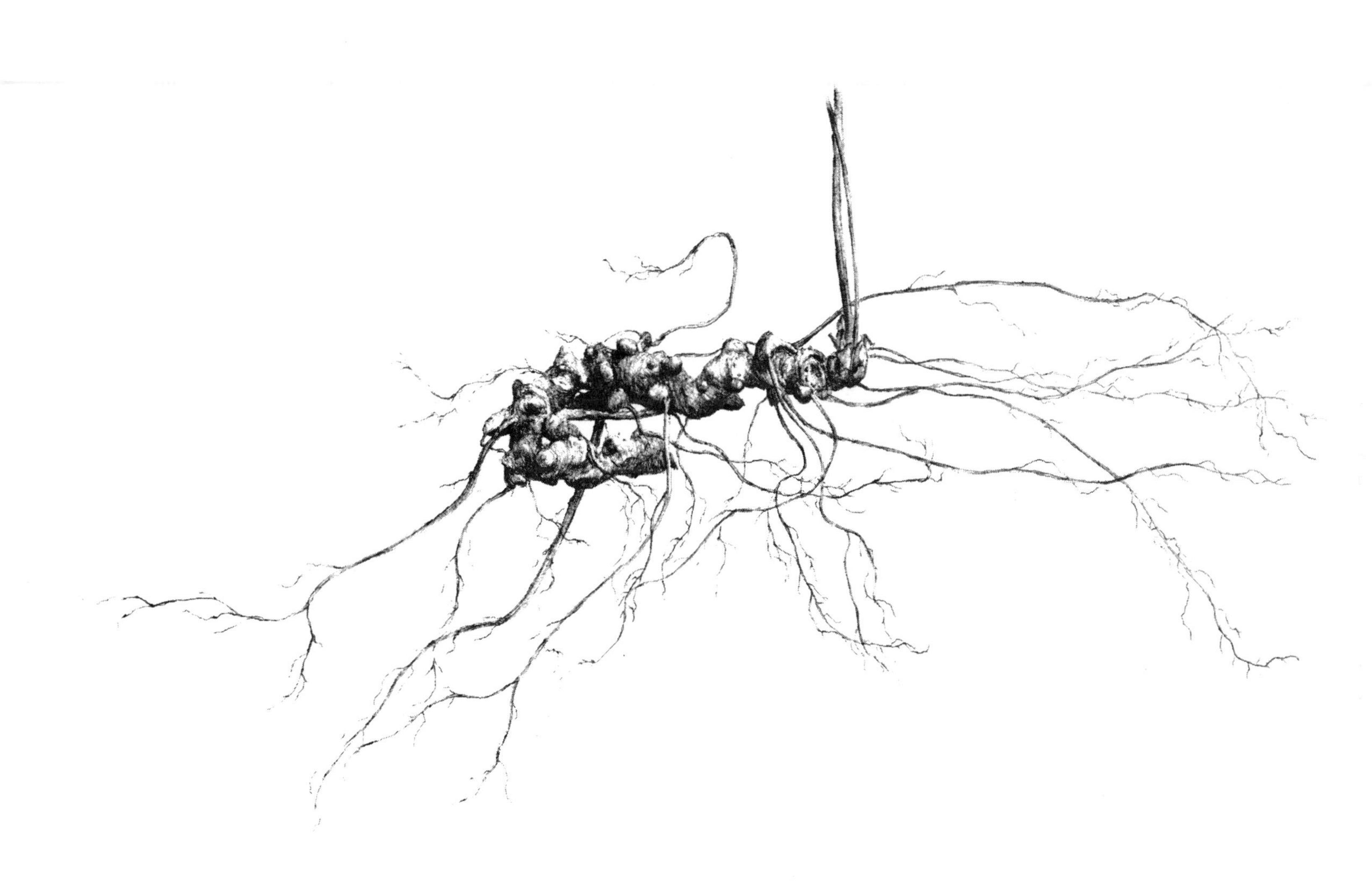

Arisaema triphyllum

Araceae, Arum family

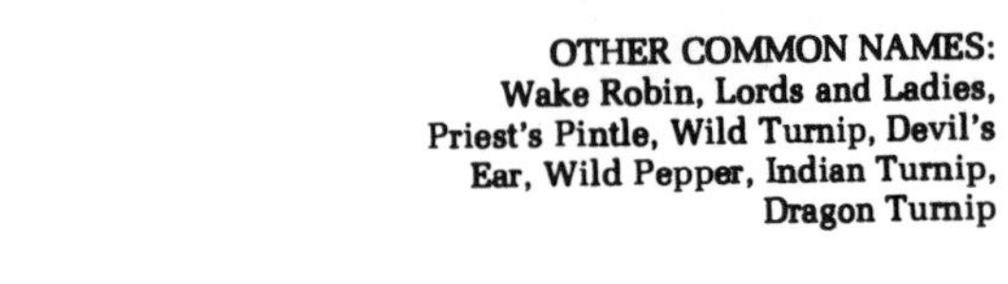

OTHER COMMON NAMES:
Wake Robin, Lords and Ladies, Priest's Pintle, Wild Turnip, Devil's Ear, Wild Pepper, Indian Turnip, Dragon Turnip

JACK-IN-THE-PULPIT

Jack-in-the-Pulpit inhabits moist woodlands throughout the Eastern half of the continent. Because of its unique appearance and wide distribution, it is one of the most known and loved of our woodland plants. Early in the spring, usually March or April, the leaves and flowers may both be seen bursting forth and unfurling in unison. The plant can attain one, to more than two feet of height. Each plant usually has two compound leaves, each leaf being divided into three leaflets which shade the flowering portion. The flowering portion is made up of a leaf-like spathe, folded over as a hood to form the "pulpit" sheltering the tube-like spadix popularly known as the "Jack." At the base of the spadix, tucked well out of sight, are the actual flowers. The color of the spathe and spadix may be anywhere from pale, washed-out green, to dark green and purple-striped. Later, a cluster of berries forms at the base of the spadix, turning bright red in autumn.

The name wake robin comes over from England where it refers to a European arum. The early spring arrival of the robin coincides somewhat with the unfurling of the plant.

The name lords and ladies must refer to the appearance of a patch of these plants resembling a group of aristocratic Victorians, each mounted on a regal pedestal and discretely shaded by striped awnings or canopies.

A less Victorian image is reflected in the name priest's pintle. The word *pintle* is old Anglo Saxon, and here refers to the phallic appearance of the spadix.

The name devil's ear takes inspiration from the elfin, pointed-eared appearance of the spathe.

The plant is perennial and arises from a corm every year. This corm is full of edible starch, frustratingly combined with intensely acrid and irritating crystals of calcium oxalate. These crystals are sharp-pointed and imbed themselves in the mucous membrane tissues, acting as an irritant both chemically and mechanically. It has been a favorite practical joke of country urchins to offer some unfortunate victim—a city slicker, talkative preacher, or over-zealous back-to-the-lander—a bite of Indian turnip.

I strongly recommend that if you think you would like to pass the brunt of this classic joke along to some new victim, you first try a bit yourself and see how much humor you find in the sensation. It is, of course, this firey hot acridity that gives names like wild pepper and dragon turnip.

Jack-in-the-Pulpit corms were often used as food by the American Indians, just as other members of the arum family are used by hunting, gathering, and simple agricultural societies in other parts of the world.

The calcium oxalate crystals will dissipate after long exposure to heat and dry air. Peter Kalm, an eighteenth century writer, tells in his *Travels In North America*, how the Indians cooked not only the corm but whole plants by placing them in a pit in the ground and building a long-lasting fire over them. "How can men have learned," he remarked, "that plants so extremely opposite to our nature are eatable and that this poison which burns on the tongue can be conquered by fire?"

It seems that heat, time, and exposure to dry air are the three main factors involved in preparation of Jack-in-the-Pulpit for eating. Various combinations and proportions of these three factors will result in an edible product. One of the best ways to prepare the corm is to slice it into very thin chips, place them in a warm, dry place and forget about them for anywhere from several weeks to more than a year, depending on the relative degree of heat and dryness. After sufficient time has passed, the chips should be tasted *cautiously*. If there is no hot, acrid or prickly sensation on the tongue or back of the throat, they can be roasted,

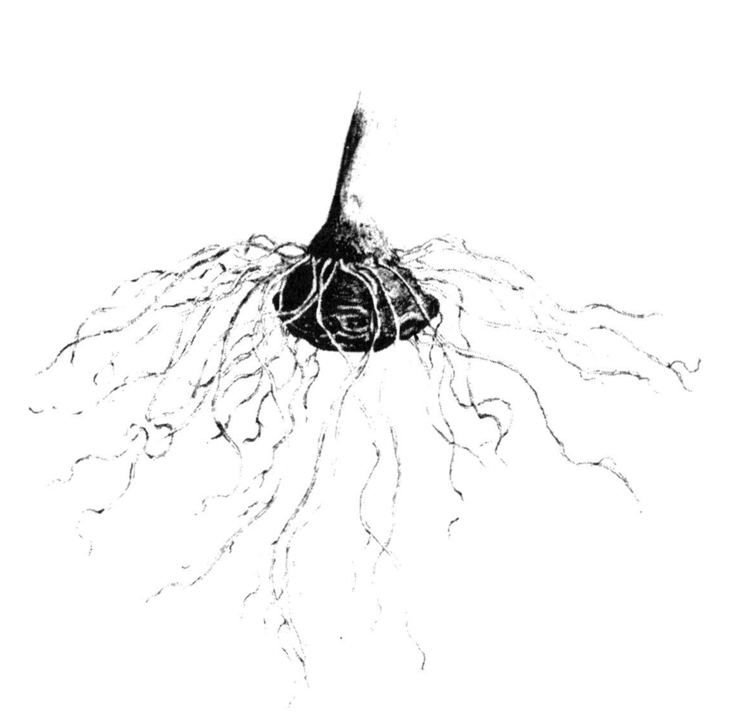

salted and served like potato chips, crumbled and made into a cooked cereal, or ground and used as a flour, a flavorful addition to other flours.

The reason I emphasize caution in the initial tasting of the Jack-in-the-Pulpit is a result of my own first experience. After waiting five months for the "Jack chips" to ready themselves, I took them down from the attic and toasted them with a little salt. After the first nibble, I promptly pronounced them delicious and started gobbling them as if I had fasted for the entire drying period. After I had eaten about 20 or 30 of the chips, the acrid burning sensation from the first chip hit me. Whereupon I realized that I would have to stomach the acridity of the rest of the chips. I am thankful that this acridity had been somewhat dispelled, so that the only consequence was a few tormented but enlightening hours, experiencing another aspect of the awesome power of Mother Nature.

Medicinally, the corm of the Jack-in-the-Pulpit was official in the United States Pharmacopea from 1820 to 1850, where it was described as a stimulant, expectorant, diaphoretic and irritant administered in a syrup or in ointment form against: colic, flatulence, asthma, whooping cough, aphthous sores and ringworm. I feel certain that it is the acridity which caused it to be stimulating and expectorant.

An Indian remedy for headache was to dust the head and temples with the powdered roots. This caused a heat which would either relieve the headache or provide a new pain to worry about.

The underground portion of the Jack-in-the-Pulpit is as distinctive as the above-ground. It consists of a shallow corm surrounded by a circular halo of smooth white roots.

Trillium erectum

Liliaceae, Lily family

OTHER COMMON NAMES:
Wake Robin, Ill-scented Wake Robin, Stinking Benjamin, Indian Shamrock, Three-leaved Nightshade, Daffy-down Dilly, Ground Lily, True Love, Nosebleed, Birthroot, Beth Root, Bettroot, Birthwort, Bath Flower, Squaw Root, Indian Balm, Mountain Lettuce, Bumble Bee Root

TRILLIUM

The Trilliums are found in rich, shady, upland woods throughout southern Canada and southward to Georgia.

Various species of Trilliums are used interchangeably in herbal medicine and usually vary only in color and the posture of the flower. Because the species most often cited in herbal manuals is *Trillium erectum*, this is the one described below. The erect flowered Trillium can be found blooming from as early as March, in the southern part of its range, through June.

The plant stands between one and two feet tall and is quite distinctive because of the way it is divided consistently into threes and multiples of threes. From the single stem radiates a whorl of three large, rounded leaves. From the center of the whorl of leaves the stem continues on to bear the flower, which in *Trillium erectum*, stands erect and can be purple, maroon, red, pink, greenish-yellow or white. There are three sepals, three petals, and usually six stamens surrounding a three-angled pistil which forms a somewhat six-lobed, dull purple or red berry about mid-summer. The plant is one of the pleasantest flowers to grace the vernal woods. Because of its early spring appearance, it shares with the Jack-in-the-Pulpit, the name wake robin.

The great diversity of common names assures us that Trilliums have been a part of the consciousness of country people and woodland wanderers for centuries.

The flower, in spite of its beautiful appearance, is also well known for its not attractive smell, as the names ill-scented wake-robin and stinking Benjamin indicate.

In reference to the unpleasant smell of the Trillium flowers, I recall a story told to me by a man in northeastern Maine who regularly serves as guide and research assistant for groups of botanists who come into his area. He recounts the rather embarrassing instance on such a botanical expedition when he was sitting over a fallen log answering the "call of nature" only to find himself face to face with one of the more near-sighted professors who thought that surely he was sniffing out the location of a bed of "stinky Benjamins."

Indian shamrock and three-leaved nightshade refer to the structure of three leaves. Daffy-down-dilly is an old poetic name for daffodils. This group of flowers was apparently broadened to include Trilliums.

The names having to do with birth, like birthroot and its corruption bethroot or bath flower, refer to the fact that the root, made into an infusion, was used by American Indians as an aid in childbirth, to stop bleeding after parturition, and as an application to sore nipples. It has been used as an astringent, a tonic to the female reproductive system, and to treat related "squaw" disorders. Its astringent properties make it useful for all kinds of bleeding, from simple nosebleed to more serious internal hemorrhaging. The powdered root, boiled in milk, is used as a remedy for diarrhea.

As an external remedy, either the root made into a poultice, or a lotion of lard or bear grease in which the leaves have been boiled, is applied to sores, tumors, indolent ulcers, skin irritations and insect stings, hence the name "Indian balm."

The leaves are called mountain lettuce in some parts of the Appalachians and are eaten either in a salad or cooked as a potherb.

The dried rhizome and roots of Trillium were official in the National Formulary from 1916 to 1949, where it was classed as an astringent, tonic, alterative and expectorant, and more recently as a uterine stimulant.

The name bumble bee root perhaps derives from the fact that it can be used to treat bee stings, but more likely it comes from the resemblance of the stubby perennial rootstock to the shape of the rotund bumble bee.

The roots are gathered in the late summer or early fall. Discrimination is advised in gathering them, as the plant is endangered and protected in some areas. It is recommended that you become familiar with *all* endangered plants in the state you are appreciating at a particular time. This will assuredly increase your sensitivity and awareness, rather than limit it.

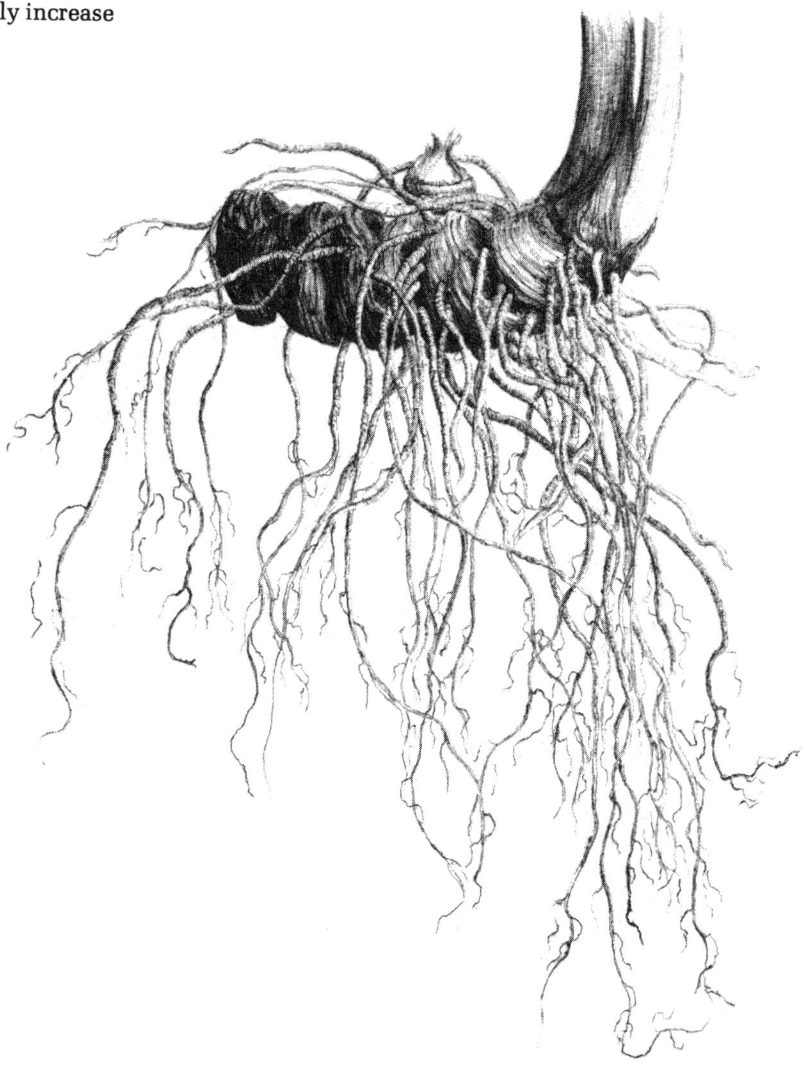

Panax quinquefolium

Araliaceae, Ginseng family

OTHER COMMON NAMES:
Sang, Redberry, Five Finger Root

GINSENG

Ginseng is native to eastern North America, from Canada south to Georgia, and in some areas of Oregon, Washington and southern British Columbia. It prefers rich, shaded, hardwood glades or coves. The soil where it grows is usually moist but well drained and not soggy.

The plant itself is not nearly as flamboyant as its reputation. It is a herbaceous perennial that rarely exceeds two feet in height and dies back every fall. It has one central stem from which two to five leafstems radiate. Each leafstem ends in a palmate cluster of usually five leaflets; hence the generic name *quinquefolium* and the common name five finger root. Arising from the center of the whorl of leaves is a cluster of five to twenty tiny greenish-yellow flowers blossoming in May or June, followed by a clump of two or three seeded berries that turn a bright shining red between August and October.

Ginseng is certainly one of the most famous herbs in the world today. In the Orient, Ginseng root is very much valued as an almost magical restorative tonic and has been used as such for thousands of years. A great deal of research, experimentation and analysis has been done on Ginseng in the Soviet Union, Germany, China and Japan.

To summarize some of the findings: Ginseng has been found to be a "non-specific" restorative and rejuvenating tonic. It helps to balance the body's metabolism, increase the body's ability to deal with stress, and through these actions stimulates both mental and physical activity, especially of the weak and aged. Ginseng has been found beneficial and stimulating to the endocrine glands which are the principle regulators of, among other things, hormone flow. It is probably this property that gives Ginseng the highly publicized and exaggerated reputation of being an aphrodisiac.

Ginseng has also been found to increase the body's resistance to adverse influences, whether chemical, biological or physical. Yet, Ginseng is *not* a cure-all. In fact, it is almost never used by itself to treat specific illnesses. Its purpose is to strengthen the system and to facilitate the action of other therapeutic agents.

In the United States, except for a brief appearance on the secondary list of the Pharmacopea from 1840 to 1870, where it was classed as a mild stimulant and stomachic, Ginseng has been all but ignored by American medicine, excluding American Indian medicine where it has been highly valued. The Cherokee word for Ginseng means "plant of life."

As with so many of God's gifts, Ginseng has also been primarily prized for the money it will bring. Even to this day, almost all the wild Ginseng gathered in this country is exported to the Orient, where it is more marketable.

The main reasons that Ginseng seems to be neglected here revolve around two key concepts: "non-specific" and "tonic." Ginseng has properties that help strengthen and protect the body against *non-specific* forms of stress, such as the chemical stress (poisons) of drugs and pollutants; biological stress such as disease germs; and against physical stress including exertion and exposure. But our medical-industrial complex has a hard time dealing with such non-specific properties. Because Ginseng acts in such an overall non-specific manner, it is difficult to measure and test, and even more difficult to stuff into a conceptual gelatin capsule labeled "cure for such-and-such."

The fact that Ginseng has been recognized as an excellent tonic is also detrimental to its popularity. Most of the older herbals and medical books describe many different substances as tonic. But it seems that in modern medical practice this term is no longer used in a general sense. When an agent is tonic, that means it imparts tone. Tone is a state of healthy elasticity or resilience when an organ or an organism's functions are in optimum balance; in tune, so to speak. Tone (perhaps it could be called health) is also a difficult condition to measure and to establish specific criteria for. Apparently, in most medical research today, the objective of maintaining good health has taken a secondary role to the development of new drugs to attack specific disease organisms. This is not to say that the development of new, specific pharmaceuticals is not a vital objective, but it is important that the limitations of such a policy be recognized.

The words of Dr. Bruce Halstead, director of World Life Research, associated with the World Health Organization, make the point.

"I've tried everywhere," he states, "I can't get any pharmaceutical company to support it (tonic research) because of the FDA's regulations which are for specifics...the western philosophy says 'Let's be specific. If you have a cough, any drug used should only stop the cough. It cannot do anything else.' The Oriental philosophy is much more integrated and broader." (6)

In the last few years, Americans have started to realize the importance of a natural diet in the maintainence of good health. There has also been an increased interest in what is going on inside the newly-opened doors of China. Along with these trends has come an increased interest in Ginseng. Ginseng can now be bought in almost any health store. It is ironic to note that it is usually much easier to obtain the lower grade of Ginseng that is cultivated in the Orient and exported to this country than it is to obtain our own native wild Ginseng, even within a few hundred miles of where it grows. Along with this upsurge in interest has occurred some research concerning Ginseng at the National Cancer Institute and at the Universities of Michigan and Washington.

Ginseng is best ingested daily in small, regular amounts over a long period. This can be done by nibbling a small quantity of the whole root or by making a tea out of the powder. I usually keep a Ginseng root on my writing desk or on the dashboard of my car when I'm traveling, and I try to remember to take a bite of it every day. I usually feel well and healthy, and don't expect to reap the full benefits of my Ginseng habit until after my 120th birthday.

The effects are subtle, general, and hardly perceivable to most people, so don't expect rushes of energy or hallucinations. Those who are most sensitive to the effects of Ginseng usually are on a pure food or vegetarian diet.

Because of the incredible demand and high price placed on Ginseng, it must be considered an endangered and valuable resource. If you should find some and want to harvest it, certain steps can be taken to gather it in a conscientious manner. It is a good policy to gather Ginseng only in the fall after the berries have ripened and can be planted to ensure future generations. It is also advisable to take only the larger, older plants, which have bigger roots. When Ginseng is grown commercially, it is considered ready to harvest only after the plants are between five and eight years old. In the wilds, it is possible to estimate the age of the Ginseng by the number of leafstalks or "prongs" on a plant. The first year the Ginseng germinates, it has only three leaflets and more or less resembles a delicate strawberry leaf. As the plant gets older, it will develop a main stem with two leaves (two "prongs") and perhaps have a flower stem between them. Usually by the third or fourth year, the plant develops into a three-pronged bunch of "seng" and starts producing berries. ("Bunch" is a mountain term referring to one single plant and probably came into use because each plant is comprised of a bunch of prongs or leaf clusters.) After about six or seven years of growth, the plant develops its fourth leaf-prong and usually produces a good crop of berries each year. I have heard stories of Ginseng plants with as many as eight prongs, but I personally have never found a "bunch" with more than five. When I counted the annual stem scars on that five-pronged plant, I found it to be twenty years old. I usually limit my harvest to Ginseng roots with four or more prongs. This way I know that the plant will have a sizeable root and will have had several years to reproduce.

Another point in favor of gathering the older plants is that as the Ginseng root ages, it gets larger and more complex and often puts out a newer auxiliary root, a sort of spare which exists to maintain and nourish the plant in case of damage to the main root system. Since "damage to the root" is a ginseng digger's middle name, this natural phenomenon can be used to the Ginseng population's advantage in the following method: The larger main root or roots are removed by cutting or twisting them off after digging very gently and cautiously around the plant. The smaller auxiliary root is left attached to the upper portion of the plant and is then replaced in the earth with as little disturbance as possible. Ideally, this whole process can be done without even removing the secondary root from the ground. When I did this in midsummer, the plant wilted at first but after a few days seemed to recover from the shock and appeared as if nothing had happened.

Another similar technique that can be used on plants that don't have auxiliary roots is to carefully replant the portion of the neck where next year's bud is forming. This isn't always successful, but the chance that it will be successful makes it worth doing. In these ways we can protect our Ginseng and eat it too.

The beautiful humanoid shape of the Ginseng's root certainly adds to its mystique and makes it one of the most intriguing of our subterranean treasures.

The name ginseng, which is supposed to be a corruption of the Chinese *Jen-shen* referring to the shape, translates as *man-root*. Manchurian myths about the origin of Ginseng say that the sacred man-root was a celestial spirit which lived in the ground and grew as a plant. After three centuries of growth, it was called by the starlight to reunite with the Spirits of the Universe. Whereupon the spirit sprang from the earth in the form of a Ginseng-man. Before ascending into the heavens, the Ginseng-man gave his white blood to humanity in the form of the Ginseng root which is a panacea for all ills and protection against mortality.

Sanguinaria canadensis

Papaveraceae, Poppy family

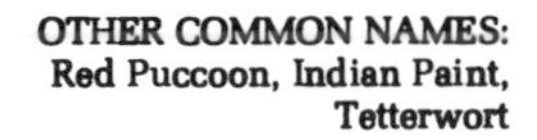

OTHER COMMON NAMES:
Red Puccoon, Indian Paint, Tetterwort

BLOOD ROOT

Blood Root is native to North America, east of the Great Plains from Canada, south to Florida. It is found in rich and usually open woodlands.

Blood Root is one of the very first plants to flower in the spring. As it first rises out of the ground, the single, small, reniform (kidney-shaped), lobed leaf is wrapped around the bud. Once out of the ground, the bud swiftly outgrows its protecting leaf. The flower bursts free of two enclosing sepals which fall away to expose eight to twelve spreading white petals. In the center of the petals is a cluster of golden stamens, usually twice as many stamens as the number of petals. The flower is followed in late spring by a slender, two-valved seed capsule containing numerous seeds. During the rest of the growing season, the leaf comes into its own and can attain a size of twelve inches across.

The root is probably one of the most dramatic that I will discuss. When fresh, it is about the size and shape of a man's finger; when cut or broken, it bleeds a red juice. American Indians used the root medicinally, and to paint their bodies for various occasions.

Calling the plant by its Indian name, *Puccoon*, John Smith wrote in 1612, "Pocones is a small root that groweth in the mountains which being dryed and beate in powder turneth red and this they use for swellings, aches, annointing their joints, painting their heads and garments." But that's not all they painted. "...and at night where his lodging is appointed, they set a woman fresh painted red with Pocones and oile, to be his bedfellow."

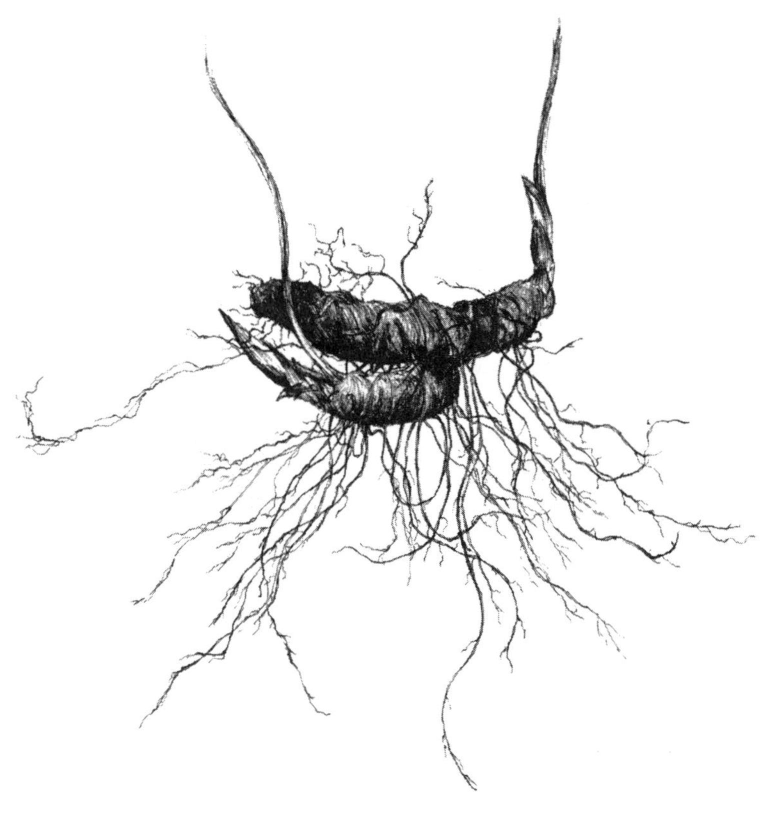

Colonel William Byrd, Virginia planter and scholar, in his *History of the Dividing Line*, (North Carolina - Virginia) wrote that such practices were still going on in 1729, much to the chagrin of the chaplain of his survey party, who "observ'd with concern that the Ruffles of Some of our Fellow Travellers were a little discolor'd with pochoon, wherewith the good Man had been told those Ladies us'd to improve their invisible charms." My research indicates that this is the first recorded case in the New World of what in modern times is known as "lipstick on the collar, telling a tale on you!"

Blood Root is still used today as a natural vegetable dye largely for wool and other textiles, now that body painting, after a brief resurgence in the 1960's, has decreased in popularity. Some excellent white oak splint basketry is still woven by the Cherokees. The splints are often dyed with Blood Root to give them a soft red color.

Medicinally, the dried rhizome of the Blood Root was official in the United States Pharmacopea from 1820 till 1926, and remained in the National Formulary from 1925 to 1965, classed as a stimulating expectorant, emetic, tonic and alterative. The plant drug, if taken internally, should be used with caution because, like many drugs, it is poisonous in all but small doses.

Externally, the powdered root or tincture acts against fungoid tumors, ringworm, warts and other skin infections. Hence, the name tetterwort. Reportedly, Blood Root tea used as a wash and taken internally has helped with poison ivy infection when nothing else would.

The powder is used as a snuff for nasal polyps. Injections of the strong tea are used to treat hemorrhoids and vaginal infections. In small doses internally as a decoction, it acts as a stimulant to the digestive organs, as an expectorant in respiratory problems and in tincture form to treat liver ailments. In larger doses it acts as an emetic.

The particular root I have drawn is a healthy one with a secondary root as an outgrowth below the first, and a growing tip at the end of each root.

Polygonatum biflorum

Liliaceae, Lily family

OTHER COMMON NAMES:
Sealwort, Conquer-john, Lady's Seal, St. Mary's Seal, Drop-berry

SOLOMON'S SEAL

Solomon's Seal grows in moist, shady woods and thickets east of the Great Plains, from southern New England south to Florida and Texas. It usually appears on the woodland wildflower scene by late April. The single stem supports a series of simple, alternately arranged, plume-like leaves along its length. There is both a large and a small variety, so the stem can be anywhere from one to six feet in length, though the plant never attains that in height, being always bent over in a graceful arc. From the axil of each leaf dangles a cluster of one to nine greenish-yellow, bell-shaped flowers. These form pea-sized, bluish-black berries in autumn which usually remain on the stem even after the leaves have dropped.

The shoots can be cooked in the spring as a green vegetable, and the flowers are delicious either steamed or added to salads. The roots as well can be boiled by themselves or added to soups and stews. I have eaten them both raw and cooked often enough and find that while they would do in an emergency, they leave a bite in the back of the mouth, which keeps me from craving them as an everyday woodland snack. European varieties of this plant have, however, been a major food source during times of famine.

Medicinally, Solomon's Seal is classed as an astringent, demulcent, expectorant and tonic. In a decoction, singly or with other herbs, it has been used for various chest and lung problems.

As a mucilagenous tonic, it is known to be healing and restorative to the alimentary canal; especially useful in stomach and bowel inflamation and piles.

Solomon's Seal is also useful for menstrual irregularities, cramps, leucorrhea and many of the other ailments classified by most early herb manuals under the broad heading of "female complaints."

John Gerarde, a 16th Century herbalist (and chauvinist), seems to use a particularly broad interpretation of this category of "ailments."

"The roots of Solomon's Seal," he writes, "stamped while it is fresh and green and applied, taketh away in one night or two at the most, any bruise, black or blew spots gotten by fals or women's wilfulness in stumbling upon their hastie husband's fists, or such like."

This brings us to the external use of the herb. (Even though family quarrels are often considered to be internal, Gerard notes that their consequences are often clearly external.) Another classic method for using Solomon's Seal as a black eye remedy is to mash the fresh roots into a poultice mixed with cream.

The only black eye I've had a chance to experiment with directly, was located on a rather willful ten year old boy. I mashed some fresh Solomon's Seal root, and in the absence of cream, used some powdered milk. The young fellow did agree to try it. He was under a bit of emotional stress, so I can't be too sure about the objectivity of his testimony, but still feel that it is only fair to report to you that he wasn't too impressed.

The extracts of the root have been used for centuries as a facial treatment to diminish freckles and lighten the skin.

The Solomon's Seal's perennial rhizome is light-colored, fleshy, and mucilagenous when fresh. The round, annual stem scars are very clear in the drawing. It can be told by counting these scars that this particular rhizome was in its sixth year when dug. The bud for the next year's stem is forming on the right. It is these round scars' resemblance to stamped-wax envelope seals that gives the plant the "seal" part of its name. One story says that Solomon, who "knew the diversities of plants and the virtues of roots," had set his seal upon it as a testimony of its value as a medicine for all humanity.

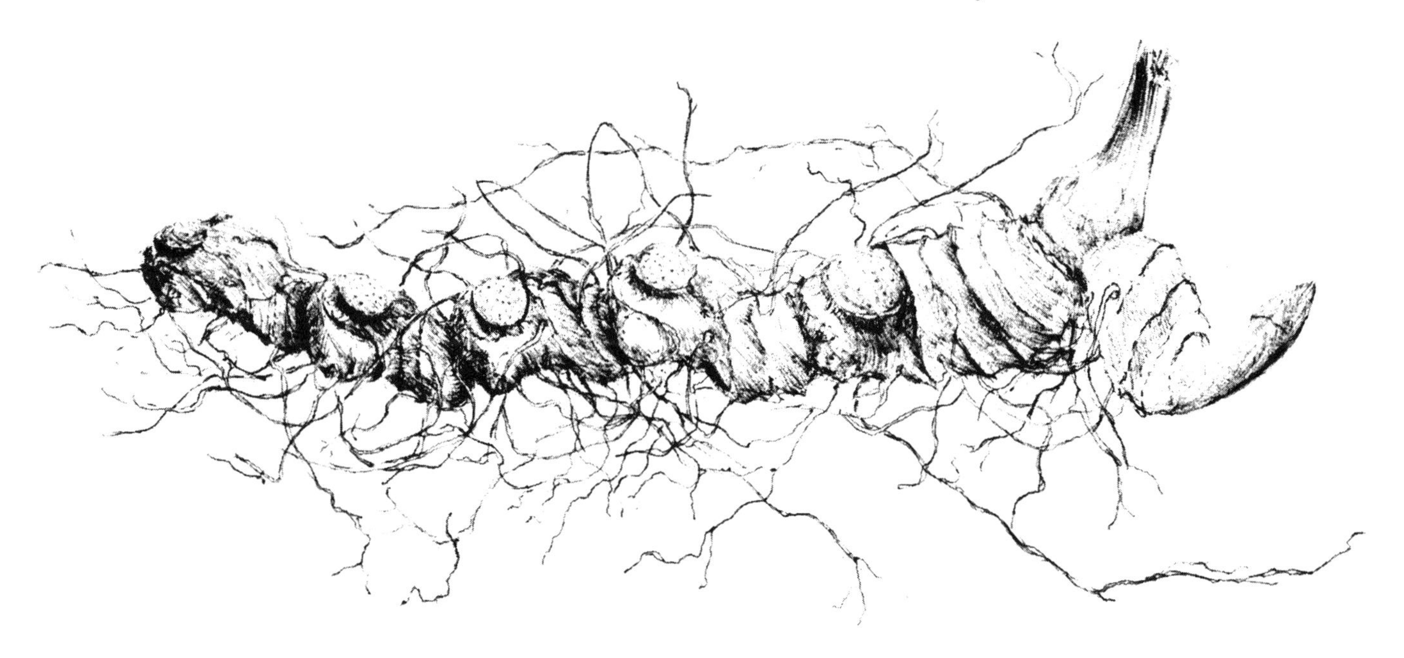

Smilacina racemosa

Liliaceae, Lily family

OTHER COMMON NAMES:
ZigZag Solomon's Seal, Scurvy Berries

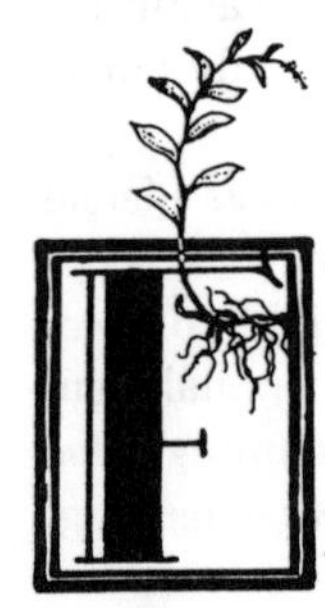

FALSE SOLOMON'S SEAL

False Solomon's Seal is found in moist, shady woods through most of the country, except for the southernmost portions. The plant is usually out of the ground by April, attains a length of up to two feet, and consists of a series of alternate, plume-like leaves. The stem angles toward each leaf and therefore takes on a slightly zigzag appearance. The leaf and stem arrangement is similar to the true Solomon's Seal *(Polygonatum)*, but the two plants can be easily differentiated by the position of the flowers. False Solomon's Seal's tiny, greenish-white flowers are in a raceme at the end of the leaf stem, as opposed to hanging along the stem from the leaf junctions. The raceme of flowers forms into a cluster of green and red speckled berries that turn bright red as they ripen in autumn. The berries are edible, the name scurvy berries suggests that they probably contain vitamin C. In early New England they were called treacle berries because their taste is indeed reminiscent of treacle, an English molasses.

It is recommended that you don't eat too many, because in large quantities they can be laxative. I have never had any adverse effects from eating a handful here and there as I'm walking through the autumn woods. However, they have never tasted special enough for me to want them in large quantities anyway. Perhaps that's another one of Mother Nature's reminders to live temperately.

In early spring the shoots are also edible as a cooked green, boiled ten minutes in salt water and served with butter, or oil and vinegar.

The rootstock was used in various ways by American Indians. Some tribes cooked them like potatoes, after first soaking them in lye to get rid of their disagreeable taste, and then parboiling them to get rid of the disagreeable properties of the lye.

Medicinally, the rootstock has been used crushed into a poultice as a wound herb, and as a tea to regulate menstrual disorders.

The horizontal perennial rhizome is somewhat slimmer than the true Solomon's Seal rhizome, but the stem scars or "seals" are still evident.

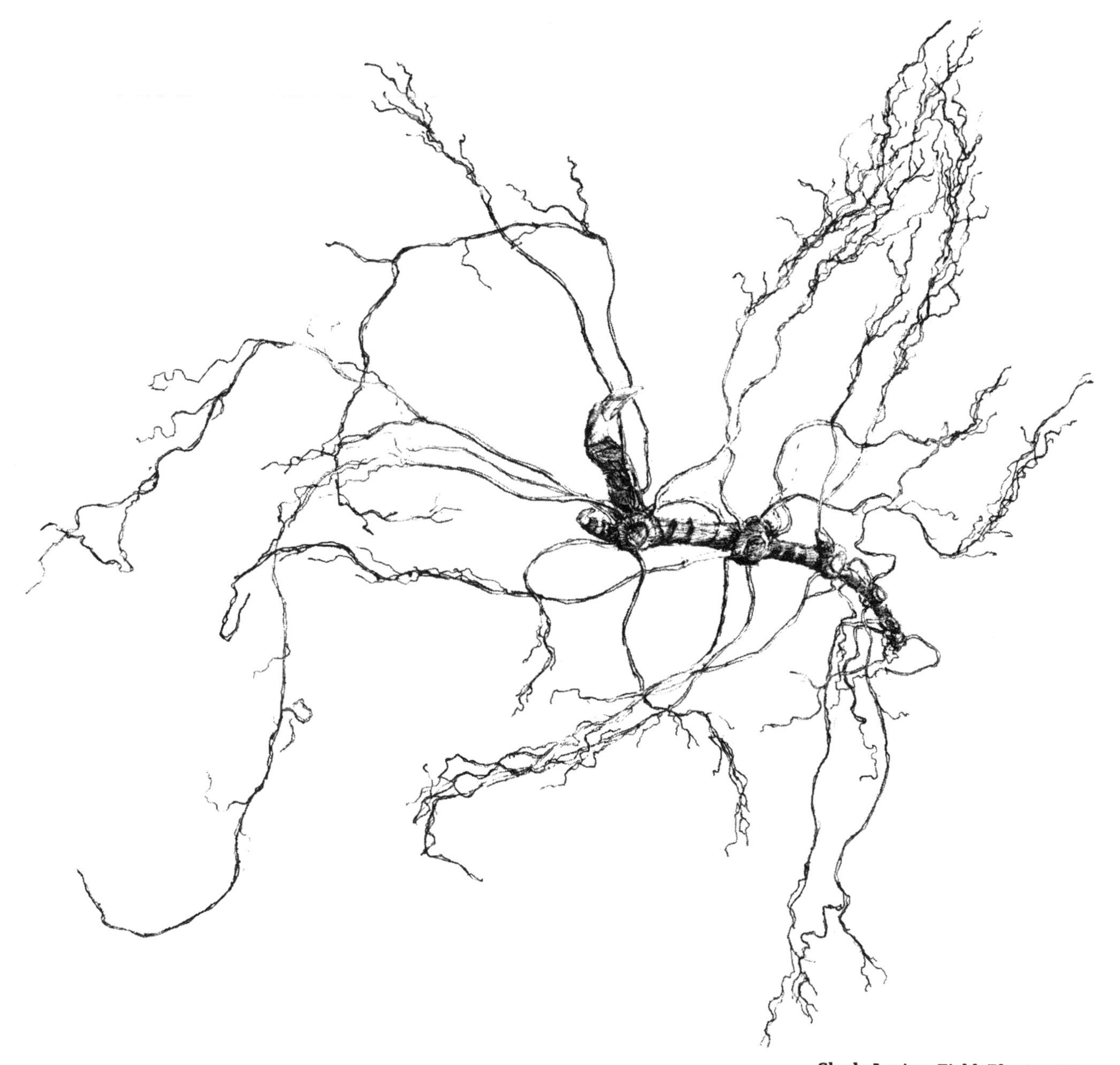

Hydrastis canadensis

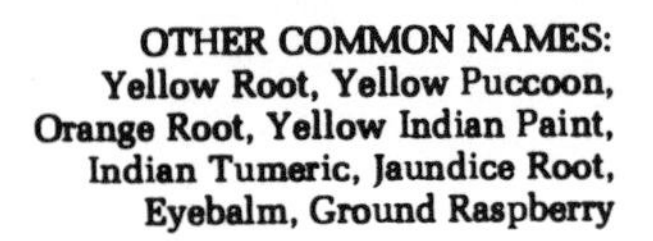
Ranunculaceae, Buttercup family

OTHER COMMON NAMES:
Yellow Root, Yellow Puccoon, Orange Root, Yellow Indian Paint, Indian Tumeric, Jaundice Root, Eyebalm, Ground Raspberry

GOLDEN SEAL

Golden Seal is a perennial plant found in rich, shady, upland woods over limestone from New York west to Minnesota and Ontario, and south to Georgia and Missouri. Because of intensive gathering, it is now rare in any of its range.

The mature plant consists of a single forked stem supporting a large leaf on one branch and a small leaf with a flower on the other branch, and is usually one foot tall. The bright yellow color of the roots extends partly up the leaf-stem to the point where it emerges from the soil. Many young woodland plants resemble immature Golden Seal. The way I confirm the identity of the live plant is by removing an inch or so of the soil at the base of the stem. If it is indeed Golden Seal, the bright yellow color lets me know.

The two leaves are palmately veined with five to nine jaggedly toothed lobes. When the flower appears in early spring, the leaves, which later attain a size of six to eight inches, are only partially developed. The upper leaf encloses and protects the flower bud. The flower is greenish-white, less than a half an inch in diameter and has, instead of petals, three petal-like sepals. These fall away as soon as the flower expands, leaving only a cluster of stamens and pistils which develop into a round berry-like cluster. The cluster ripens in August to resemble a single large raspberry from which the common name ground raspberry is derived. In a bed of Golden Seal, there are often large numbers of immature, sterile plants having only a stem with one leaf. Only after at least three years is the plant old enough to flower and reproduce.

In the past, Golden Seal has been used to make a beautiful yellow dye. This is where it gets the names like yellow Indian paint, Indian tumeric, and Ohio curcuma. In more recent times, Golden Seal has become too rare and expensive to use as a dyestuff.

Even if rare, Golden Seal is still one of the most popular medicinal herbs used today. Externally, the powdered root is used directly in cuts and small wounds as an antiseptic aid to healing and to arrest bleeding. The powder can also be snuffed for nasal inflamation and can be applied directly as a medication for sore throat, either by dropping a bit of the powder on the back of the throat or by blowing it through a straw. The strong decoction is used as a wash for festering sores, skin erruptions and eye infections. It can be used in tooth powder and will help to prevent or relieve gum infections.

I made up a homemade toothpowder by first mixing powdered Golden Seal, myrrh, and sea salt to act as disinfectants and medications. To this I added an equal amount of powdered chicory charcoal. The charcoal is gently abrasive and helps whiten the teeth and freshen the breath. I obtained the charcoal by forgetting to check come chicory roots I had been roasting in the oven. To these ingredients I added some powdered peppermint in hopes that this might help counteract the bitter taste of the Golden Seal and give the powder a refreshing toothpaste-like flavor. (It did neither.) But I found that by cultivating a proper attitude, I could really enjoy using my natural, herbal dentifrice. It does seem to whiten the teeth and even seemed to stop the case of bleeding gums I had before I started using it. The charcoal, combined with the bitter properties of the Golden Seal powder, leaves the mouth with a fresh, clean taste. But, as I said before, it *is* necessary to have the proper attitude. That bitter taste is a bit difficult for many people to overcome. However, when I am brushing my teeth at the side of a bubbling, mountain stream, bitter taste or not, it seems like just the right thing to use.

It is quite a different case in a modern bathroom, too early on a Monday morning. To look in the mirror, when you're used to the white, foamy, sweet tasting, commercial toothpaste, and see that black, gritty stuff come oozing out of the corner of your mouth and down your chin to splatter all over the sparkling white porcelain sink--well, most people find this a bit disconcerting. So alas, my inspired creation, now known as "The Bitter Black Drool," has come into a period of disuse even among my most open minded friends. As for myself, even my proper attitude has waned a bit, and now I save it for those bubbling mountain streams, or at least for situations where there is no mirror or porcelain sink.

Internally, the decoction of Golden Seal is very well known as a stomach medicine. It acts as a tonic and a stimulant to the digestive system, increasing both the appetite and gastric secretions. It is useful in poor digestion caused by weak, debilitated, or inflamed stomach or intestinal lining. It helps stimulate poor circulation in cold extremities, especially when combined with cayenne pepper. It is useful either alone or combined with other herbs, such as myrrh or slippery elm, as an injection to treat various genito-urinary infections.

For bad colds and flu, Golden Seal tea is one of my favorite remedies. It seems to be antibiotic in action. The only complaint most Golden Seal users have about the tea is its bad taste. Several herby friends of mine discovered (and I have to agree) that a little honey and a drop or two of brandy in the tea does wonders, not only to cut the bad taste but also to help warm the insides and clear the head.

An extract of Golden Seal, the drug called "hydrastine," has been used to treat malaria, and is considered second only to quinine.

Golden Seal and/or its derivatives were official in the United States Pharmacopea intermittently between 1831 and 1936, and were listed in the National Formulary until 1960.

The underground rootstock, when fresh and clean, is a brilliant yellow that can hardly be surpassed by the blossoms of any wild flower. Like many rhizomes, it grows along horizontally a few inches under the ground. If you are gathering it, the budding end portion of the older rhizomes should be broken off and replanted as a conservation measure.

Golden Seal has been in the news recently. Apparently heroin addicts found that if they take Golden Seal it will mask the presence of heroin or morphine in the urine tests that were being used by drug detoxification programs. This result could be due to a chemical reaction, but some herbalists contend that it is due to Golden Seal's strength and effectiveness as a blood cleanser and detoxifier. It seems that whatever the reason for this action, new heroin tests have been developed which are not influenced by this root.

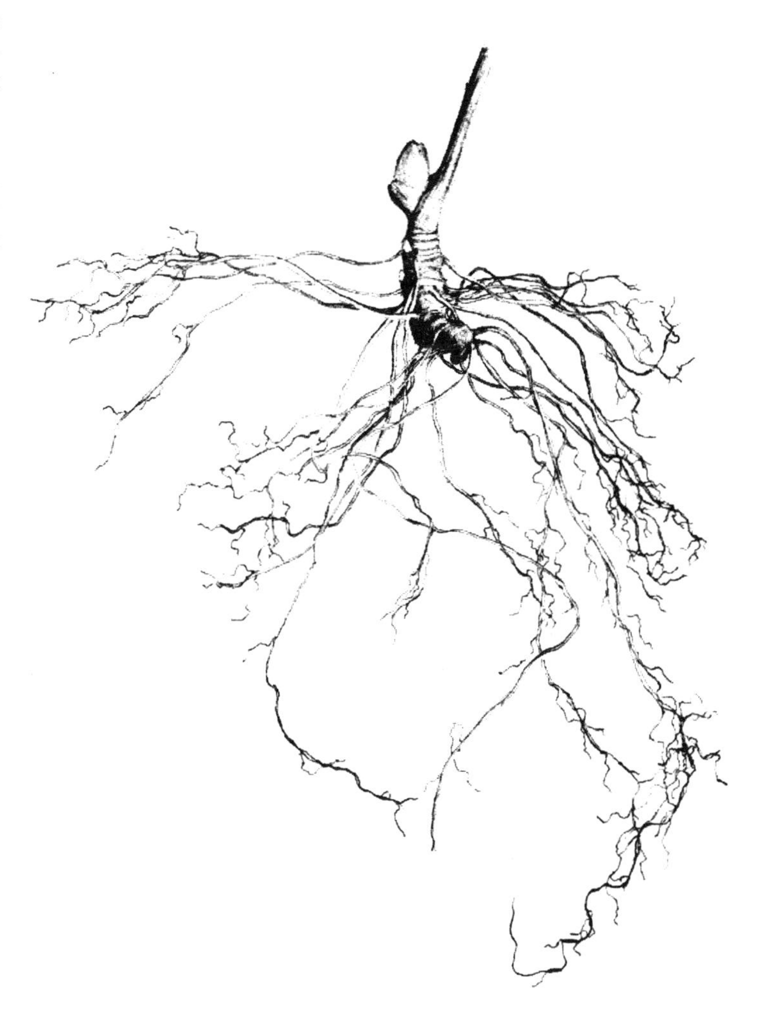

Xanthorhiza simplicissima

Ranunculaceae, Buttercup family

OTHER COMMON NAMES:
Shrub Yellow Root, Parsley-leaved Yellow Root

YELLOW ROOT

Yellow Root usually grows near wooded stream banks. It is most widespread in the southern Appalachians and the piedmont from Virginia to Florida and west to Kentucky, but is also reported as far north as southwestern New York state.

Yellow Root is a perennial shrub that grows about two feet tall. The large, much-divided, compound leaves are borne in a cluster on top of a spindly, usually single stem. The flowers dangle from several drooping spikes at the base of the leaf cluster. They are inconspicuous and totally lack petals. The five-pointed cluster of sepals is usually maroon to brownish purple in color and is in bloom in April or May. It produces seeds in June.

As a southern folk remedy, Yellow Root has been widely used and is still highly regarded in many areas. It was official in the United States Pharmacopea from 1820 to 1882. Yet it has not received much attention in the herb books so popular these days. Many people who do use it, including myself, chew a section of the bitter root regularly as a general tonic with an especially beneficial effect on the gastric system.

The flavor of Yellow Root is a very strong but "clean" bitter, with no acridity or astringency; just pure bitter! The taste for bitters is an acquired one and worthwhile to cultivate. In many civilizations it has long been customary to take bitters before meals, as a digestive stimulant. Scientists tell us that we only have four kinds of taste receptors on our tongue. They are: sweet, sour, salty and bitter. Accordingly, everything we taste is one, or a

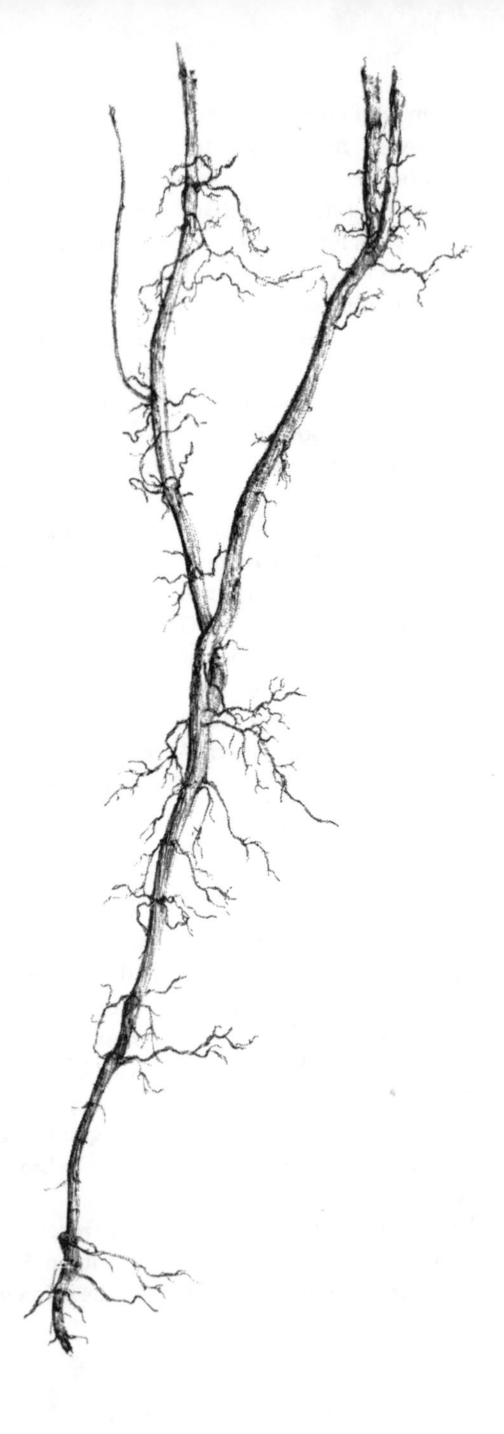

combination of these four tastes. If we avoid anything that is bitter, we deprive ourselves of a full quarter of our sense of taste! But all it takes is a little conscious effort and the proper attitude to avoid such deprivation. Actually, acquiring a taste for the bitter enhances our appreciation of other tastes. The bitter of Yellow Root not only tones the system and stimulates the flow of gastric juices, but also cleans and freshens the mouth and sharpens the taste buds. The following story will illustrate.

One Sunday, during the height of the fall foliage, several of us were in the North Carolina mountains, hiking down a well trodden path through the woods to view a beautiful, but on this particular Sunday, very touristed waterfall. The path, strewn with litter, nagging parents, and tired children did much to mar the spirit of the day. As we approached the river, I noticed beds of Yellow Root. More for distraction than anything else, I pulled up one of the plants and started to chew on a piece of the root. Feeling the intensity of the penetrating bitter juice, I offered pieces of the root to my friends, extolling its virtues and suggesting that such was compatible with the attitude of the moment. While the experience did not immediately improve our outlook, the rugged splendor of the waterfalls eventually helped us transcend the bad taste of the Yellow Root and the Throw-Away Society.

Best of all, when we stopped at a drinking fountain, everyone was overcome by the incredibly fine taste of ordinary water. More than likely, the bitter Yellow Root had so sensitized our taste buds that we could properly appreciate the quality of mountain water. Next I passed around some wild apples, and presto: instant euphoria! They were just average apples but went over like Ambrosia. Thus did the bitter enhance the sweet.

The root can also be made into a tea to be used as a mouthwash for a sore and ulcerated mouth, and as a medicine for colds and disorders of the stomach and throat. Indians have used the root for similar purposes as well as for an aid in childbirth, and as a source of yellow dye.

Because of the name, many people confuse Yellow Root with golden seal *(Hydrastis canadensis)* and use it for the same purposes with similar results. Yellow Root does in fact have many of the same properties as golden seal and some of the same chemical compounds. I would encourage this substitution because Yellow Root is much less endangered than golden seal.

Both the wood and the root of Yellow Root have the bright yellow color from which it gets its name. The root is long, slender and extends fairly deeply into the ground where it can keep its "feet" wet. It frequently branches to form new, above-ground parts. In this manner it forms extensive colonies. The roots can be gathered at any time during the year.

Cimicifuga racemosa

Ranunculaceae, **Buttercup family**

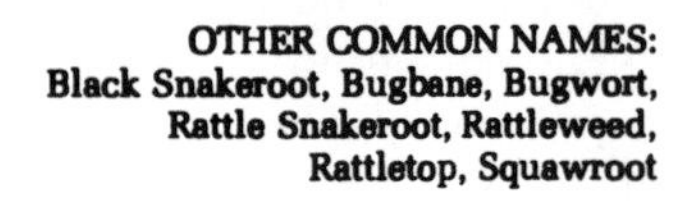

OTHER COMMON NAMES:
Black Snakeroot, Bugbane, Bugwort, Rattle Snakeroot, Rattleweed, Rattletop, Squawroot

BLACK COHOSH

Black Cohosh prefers the shady, rich soil of hardwood forests from Maine to Wisconsin, and south along the Appalachians to Georgia, and west to Missouri.

Rising three to eight feet, the spikes of the delicate, showy flowers are conspicuous as they tower over most of the understory plants of the early summer woods. The tall, spreading, compound leaf-stem is divided and subdivided into threes.

As attractive as the tall spike of flowers is to the eye, it is somewhat less attractive to the nose. Because of its peculiar smell, it has a reputation as an insect repellent which accounts for some of its names like bugbane and bugwort. In late fall and early winter the seed pods, still attached to dead stems, are full of hard, dry seeds. The rattling of the seeds in the pods as they are brushed or as the wind passes over them gives rise to the names rattleweed, rattletop, or rattle snakeroot.

Both the white men and the American Indians—specifically the Winnebagos, Dakotas, and the Penobscots—have regarded Black Cohosh as a treatment, not only for snakebite (hence the name snakeroot), but also for diarrhea, deep chest afflictions, spasmodic cough, and menstrual irregularities (which gave it the name squawroot; although it was also because the squaws were often seen digging the roots).

In herbal medicine it is considered to be astringent, diuretic, expectorant, sedative, slightly narcotic, antispasmodic, and an emmenagogue. It was recognized as official in the United States Pharmacopea from 1820 to 1936, and in the National Formulary from 1936 to 1950.

The Algonquins called the root *cuski*, meaning rough, from the appearance of the rootstock. The white man supposedly twisted the word cuski into cohosh, prefixing it with black because of the roots' dark color.

The rootstock is coarse, gnarled and irregular. The numerous scars and stumps on the upper surface are the remains of former leaf stems. When the root is fresh, as in the illustration, the young, pale, slightly pinkish buds which are to furnish the next season's growth can be seen along the portion of the root that is the base of the present year's stem. The long, fleshy rootlets produced at the lower portion of the rootstock become very brittle upon drying. They are best gathered in the fall.

Caulophyllum thalictroides

Berberidaceae, Barberry family

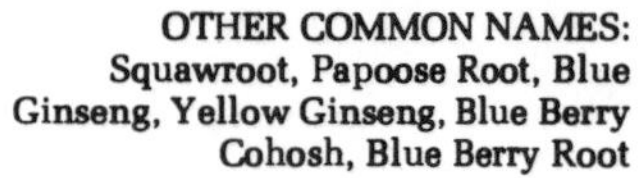

OTHER COMMON NAMES:
Squawroot, Papoose Root, Blue Ginseng, Yellow Ginseng, Blue Berry Cohosh, Blue Berry Root

BLUE COHOSH

Blue Cohosh is found from New England south to the Appalachian regions of the Carolinas. It grows in moist, mountain glades under hardwoods.

This plant usually attains a height of three feet with one or two large, thrice divided, compound leaves with many round-lobed leaflets resembling meadow rue (*Thalictrum dioicum*) only proportionately larger. The flowers are tiny and inconspicuous, with greenish or maroon-tinged, petal-like sepals. The fruit is a blue berry the size of a large pea. The berry often hangs on into autumn, long after the leaves have withered away. The blue color of these berries gives the plant so many of its common names.

As the name squawroot indicates, its application is for the benefit of women. It has been considered by Indians and white folk alike to be an excellent regulator, useful in cramps and menstrual difficulties of all kinds. It is used during pregnancy and childbirth as a uterine toner, to ease the pain of labor, and promote prompt delivery. A tea of Blue Cohosh was taken during the last three or four weeks of pregnancy.

Not limited to squaws, papoose root was also given to children to relieve cramps and colic.

Blue Cohosh has been used as a treatment for heart palpitations, high blood pressure, whooping cough and epilepsy. It is known to contain a large number of vital minerals.

It was listed in the United States Pharmacopea from 1882 until 1905, and in the National Formulary from 1916 to 1950, where it was considered to be an antispasmodic, emmenagogue, and diuretic.

The Blue Cohosh rhizome, due to its intense tangle of roots, is a little more difficult to unearth and wash than the average woodland root. When fresh, the rhizome and roots are a light tannish-yellow in color (hence the name yellow ginseng) and the next year's stem buds are tinged with lavender.

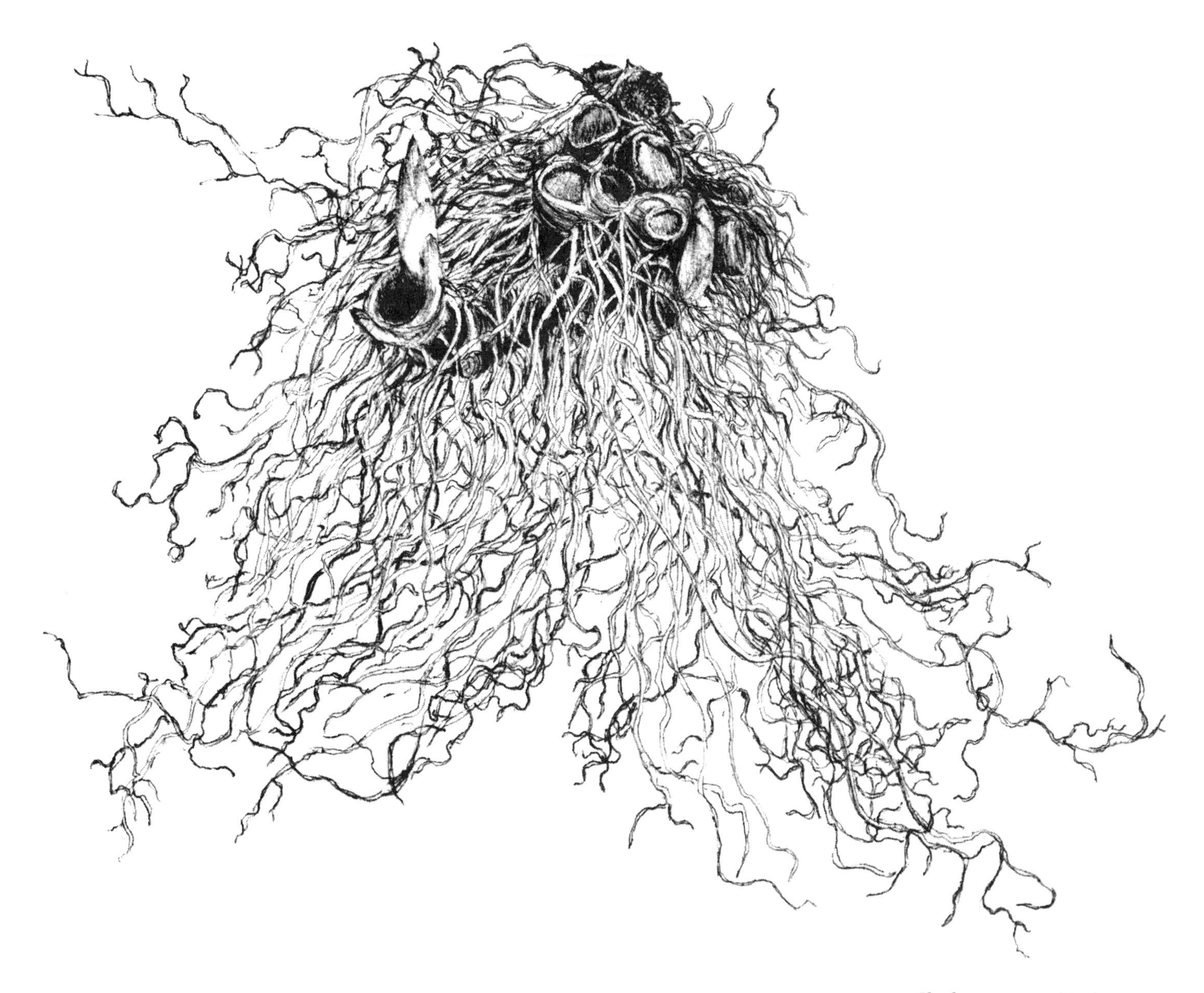

Podophyllum peltatum

Berberidaceae, Barberry family

OTHER COMMON NAMES:
American Mandrake, Umbrella Plant, Duck's Foot, Devil's Apple, Raccoon Berry, Ground Lemon, Hog Apple, Yellow Berry, Vegetable Calomel, Vegetable Mercury, Wild Jalap

MAY APPLE

May Apple is found growing in beds, in moist, rich soil either in shady woodlands or sometimes at the edge of meadows. It ranges east of the prairies from Minnesota to Texas and Florida, and in Southern Canada from western Quebec to Ontario.

In March, April, or sometimes as late as May, as the name indicates, the small, folded, fist-like ball of leaves of the May Apple pushes through the leaf litter. One or two leaves expand into a rounded, deeply lobed, umbrella shape; some leaves attaining a size of nearly twelve inches in diameter. The plant grows to a height of one or two feet. Younger plants bear only one leaf, while mature specimens have two leaves springing from the single forked stem. At the junction of the two leaf-stems a delicate, waxy, white flower is borne. By late summer the flower forms a yellowish, lemon-shaped "apple" — more correctly a berry. Usually, the May Apple plant has completely exhausted itself on the effort to reproduce, and has collapsed by the time the fruit ripens, so that the soft, yellowish apples must be picked up from the ground.

These fruit have a subtle, quite indescribable flavor which Euell Gibbons calls "hauntingly delicious." The other parts of the plant, however, should not be taken lightly. The rootstock is a very strong drug known mainly as a cathartic in tiny doses. In large doses, it has killed people. It was first listed in the United States Pharmacopea in 1820, and it remains there today (although it was dropped from 1942-1955). It has distinct medical applications but is rather drastic in action, and it should not be used as a casual home remedy.

The rootstock is considered a hydragogue cathartic, used against constipation to increase intestinal action and secretions and to produce a copious, watery stool. It is also used to increase bile flow and to stimulate a torpid liver. At one time, mercury compounds were used with similar cathartic effects in treating ailments of the liver and other organs. Thus, May Apple root came to be called vegetable mercury or vegetable calomel.

When used as a cathartic to treat constipation, it is usually combined with a more gentle laxative like senna or licorice root to modify its action.

The resin extracted from May Apple rootstock, called *podophyllin*, is used as a caustic applied to remove certain kinds of warts.

One of the Indian uses could be applied by organic gardeners today. The Menominees boiled the whole plant to make a decoction which they used as an insecticide on their potato plants.

Chamaelirium luteum

Liliaceae, Lily family

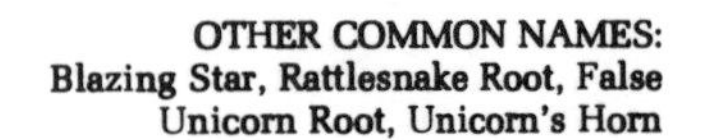

OTHER COMMON NAMES:
Blazing Star, Rattlesnake Root, False Unicorn Root, Unicorn's Horn

DEVIL'S BIT

Devil's Bit is found from Massachusetts to Michigan, and south to Florida and Arkansas in rich, moist, open woodlands.

The plant arises from a cluster of somewhat fleshy, oblong, basal leaves. The leaves are up to eight inches long at the base, diminishing in size on the upper portion of the stem. The erect flowering stem usually bears only male or only female flowers. The female plant may be more than three feet tall and is leafier than the male plant, which attains a height of only one to two and a half feet. In an average population the male plants usually outnumber the females.

The white, starry flowers bloom between May and July. Male flowers are borne on a graceful, drooping, plume-like spike (hence the name drooping starwort), and those of the female on an erect spike. The seeds are formed in an oblong capsule in late fall.

A concoction of Devil's Bit is considered to be a uterine tonic used to prevent miscarriage, to correct various forms of pelvic or uterine atony, and to ease the problems of menopause. It is also a tonic in the more general sense. Acting favorably on other parts of the genito-urinary system, Devil's Bit has been used to treat male impotence and to exert a beneficial influence in gastro-intestinal disorders. As a vermifuge it is said to expell tapeworms, but because of its strength some authorities advise against the use of Devil's Bit as a casual home remedy.

The rhizome is rather distinctive. When fresh, it is light colored and usually curves upward at the end to resemble a stubby horn, hence the name unicorn. The more recent rootlets that radiate from the central rootstock are light colored and have a soft outer covering. The older rootlets appear tough and wiry because this covering has worn away. Toward the end of the rootstock, wherever these rootlets have broken off, the rhizome has a small round hole that makes it appear wormeaten.

The name Devil's Bit derives from the bitten-off, knub-like appearance of the rhizome. The name, along with a legend, came from Europe where it referred to a European plant with a similar rootstock. The long, trailing roots of this plant, so the legend goes, possessed not only extremely beneficial healing properties, but extraordinary magic. So beneficial was it for the people who used it that the Devil himself became angered and tried to change the qualities of the root from good to bad. The power and goodness of this plant was so strong, however, that his attempts were always thwarted. Finally, the Devil flew into a rage and personally bit off every one of the roots. His rage was so searing that to this day the roots have not been able to grow back. But the remaining stub is still imbued with good medicine, and every spring it is able to put forth the tall spike of beautiful, blazing-star blossoms as a reminder that the power of goodness can always avert the forces of evil.

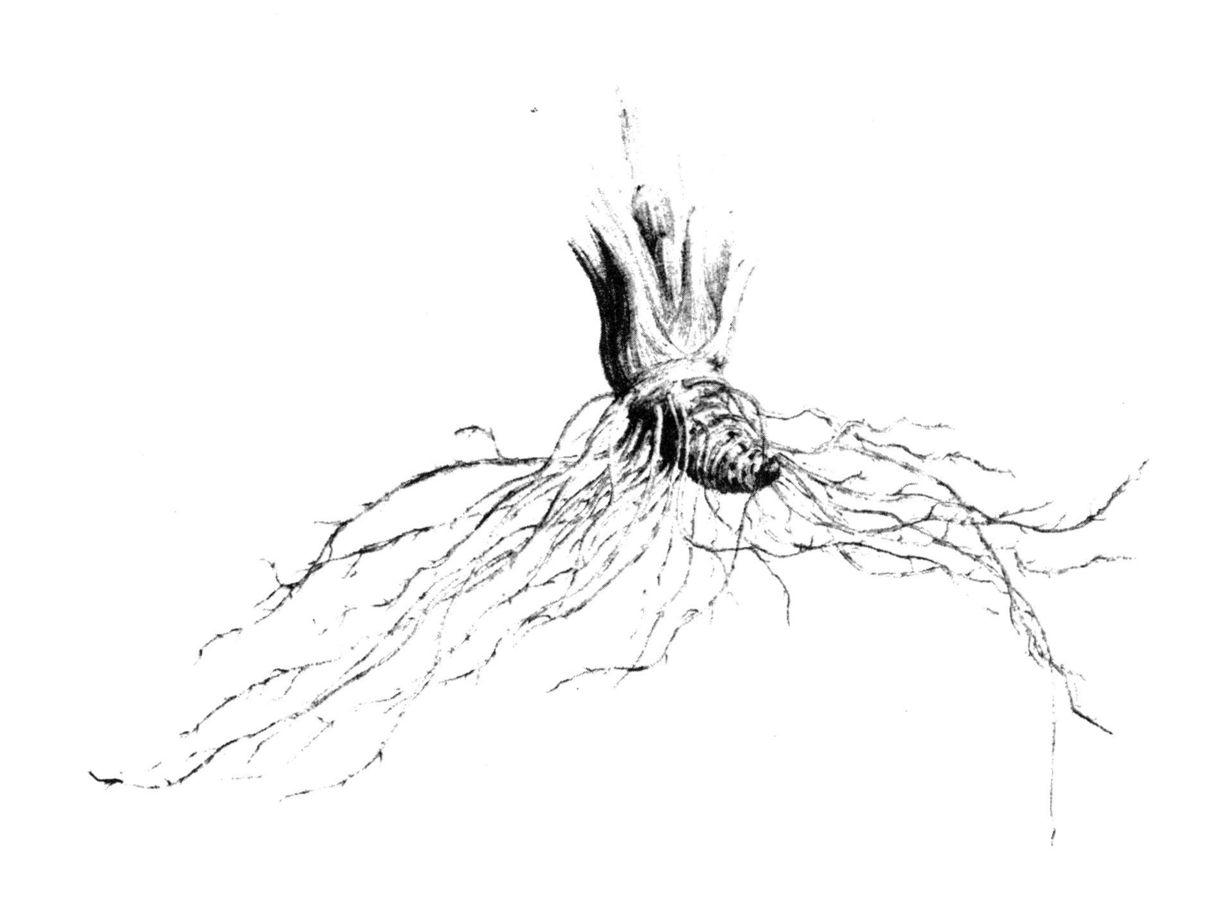

Medeola virginiana

Liliaceae, Lily family

INDIAN CUCUMBER

Indian Cucumber is found in openings of rich, slightly acid woods from the southeastern half of Canada south to Florida and Tennessee. The single stalk of this distinctive perennial plant can attain a height of two feet. A little more than halfway up the stalk radiates a whorl of five to nine, three or four-inch, stemless, elongated leaves. Above this whorl the stalk continues on to bear another whorl of usually three, much shorter, leaves. From these upper leaves springs a cluster of three to eight nodding, greenish-yellow or straw-colored flowers, usually in bloom between April and June. Come September and October it produces bluish-black, six-seeded berries. Often, in a bed of Indian Cucumbers, there are numbers of younger or weak plants that do not flower and consist of only the stem and the one whorl of leaves.

The part of the Indian Cucumber that we are most concerned with here is the crisp, white, horizontal rhizome. The plant gets its name from the fact that this rhizome, during a large part of the year, does taste much like a cucumber. Even when it doesn't, it retains its crisp, cool texture, which makes it one of the most pleasant, on-the-spot, wildwood nibbles in our forests.

Indian Cucumbers are a good addition to salads and steamed vegetables. But they can also be made into some of the nicest pickles you'll ever eat. Just use your favorite pickle recipe.

The author doesn't mind boasting that one of the high points of his several years in northern New England was winning a blue ribbon for Indian Cucumber pickles at the Fryeburg Fair in western Maine.

However, in harvesting Indian Cucumber great restraint should be exercised. They should only be taken from areas where they grow in abundance, otherwise that jar of pickles will be sitting on a museum shelf instead of your pantry. One botany manual recommends taking up an Indian Cucumber only if there are three others in a 3-ft radius.

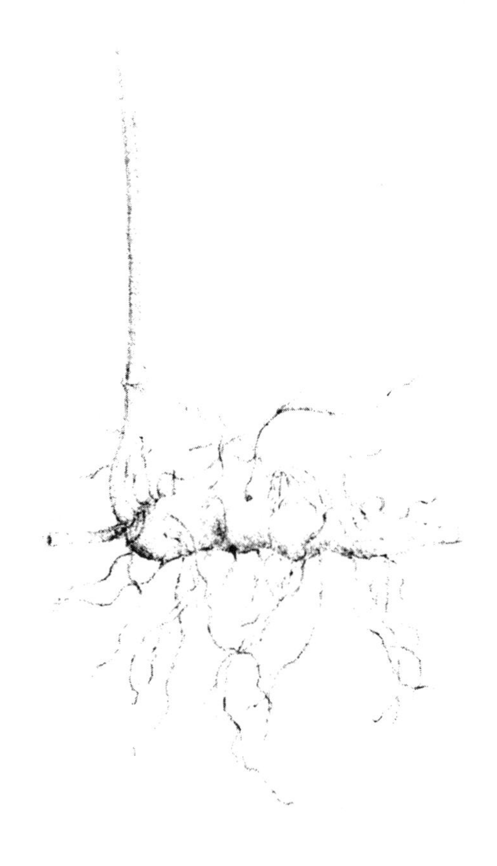

Osmorhiza longistylis

Umbelliferae, Parsley family

OTHER COMMON NAMES:
Anise Root, Sweet Chervil, Sweet Javril, Cicely Root

SWEET CICELY

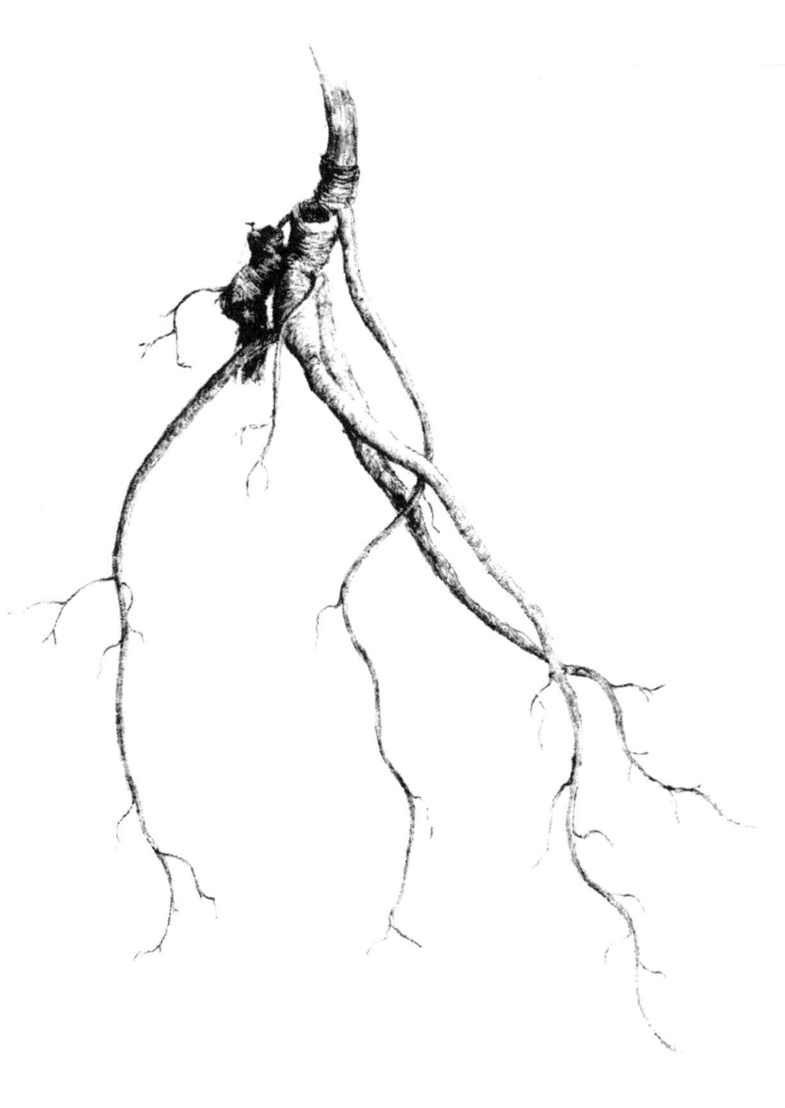

Sweet Cicely is found in the rich soil of shady woods from Nova Scotia to Ontario, south to the Carolinas and Alabama and west into Colorado.

Sweet Cicely is found blooming with tiny, white flowers in late spring. The plant stands about two feet high and bears several large, compound, toothed leaves resembling a large version of garden chervil or carrot tops.

There are two species of *Osmorhiza* in our area. Both have aromatic roots, but only *longistylis* has the delicious anise or licorice flavor. The easiest way to tell them apart, short of digging up the root, is to examine the plant when the seed is present. The seed of the *O. claytonia*, is shorter and ends in a single point. The seed of the one we are interested in, *O. longistylis*, is larger, longer, and ends in two points.

Usually by late summer most of the leaves have fallen, but the plants can still be recognized by the seeds that remain. The aromatic characteristics of the root make it useful as a carminative, expectorant and stomachic, good for soothing and settling the stomach. It can be taken as a tea or used as a flavoring for desserts. My favorite way of consuming Sweet Cicely is to eat the root fresh out of the ground, after washing it in a mountain spring. The delicate licorice flavor leaves a fresh, sweet taste in the mouth. I never cease to marvel at how pleasant our woodlands are, yielding a root that is so sweet and candy-like in its raw and unaltered state.

Collinsonia canadensis

Labiatae, Mint family

OTHER COMMON NAMES:
Citronella, Horseweed, Ox Balm,
Stone Root, Knob Root

HORSE BALM

Horse Balm is found in moist woods, often at the edges of clearings or woodland roads, from southern Canada south to Florida and west to Arkansas.

It reaches a height of two to as much as five feet. Like most members of the Mint family, it has a square stem. The oppositely situated leaves are large and rounded with toothed edges and come to a point. The flowers that bloom from July through September are very fragrant, smelling like lemon or citronella, and are borne on heads or panicles. These flowers have a pale yellow, funnel-shaped corolla with two protruding stamens bearing red anthers.

The leafy part of the plant has been used in the form of a poultice as an application to bruises, sprains and other sores. The large knobby rootstock is quite distinctive and is very hard whether fresh or dried (hence stone root and knob root).

Medicinally, the root is classed as an antispasmodic, astringent, tonic, diaphoretic and diuretic. As a diuretic, the decoction of the fresh root has been given in cases of bladder inflamation and other genito-urinary disorders, as well as for hemorrhoids and rectal problems. It is also used to treat colic, indigestion, cramps and chronic bronchitis. A decoction or a tincture taken with honey as a gargle is considered to be a good remedy for hoarseness.

The names Horse Balm, horseweed, and ox balm probably originate from the fact that this plant has been used in some of the above ways as a veterinary medicine. If your ox ever needs a good balming, you now know what to use.

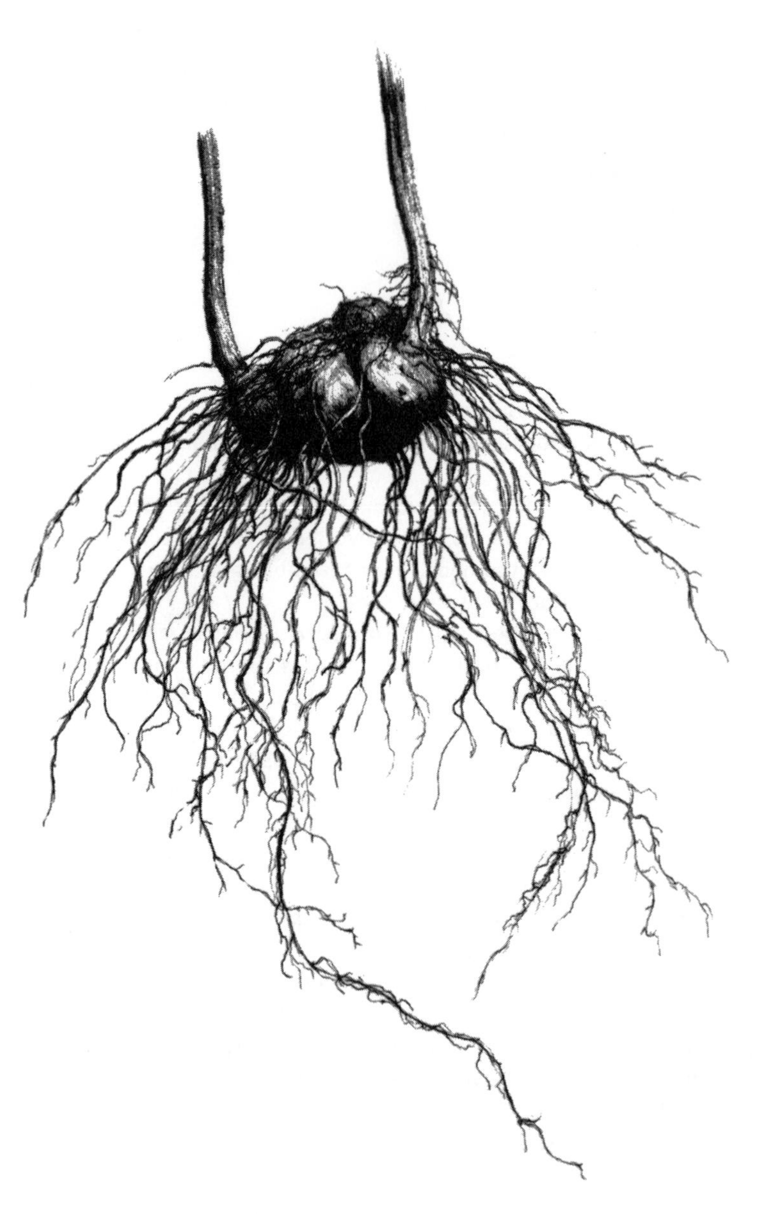

Claytonia virginica

Portulacaceae, Purslane family

OTHER COMMON NAMES:
Fairy Spuds, Good-morning-Spring, Wild Potatoes

SPRING BEAUTY

Spring Beauties are among the first to announce Spring's arrival. They are very low-growing and delicate, with slender leaves and tiny, light pink to white, five-petaled flowers intricately veined with darker pink. They blossom in March and April and have usually died back by summer.

There are at least three species of Claytonia found in moist woodlands throughout most of this country. The only obvious variation among them is the size and shape of the leaves.

The plants are perennial and spring forth each year from small roundish corms. These corms are tiny, rarely as large as an inch in diameter, and seldom grow abundantly enough to warrant the effort of uprooting a quantity sufficient for a meal. However, if you should find a large bed of them growing thickly enough so that a little vigorous digging won't wipe them out, you are in for a treat. Dig them, wash well, and then boil them for about ten minutes. They are similar to regular potatoes, only somewhat sweeter, recalling the flavor of chesnuts.

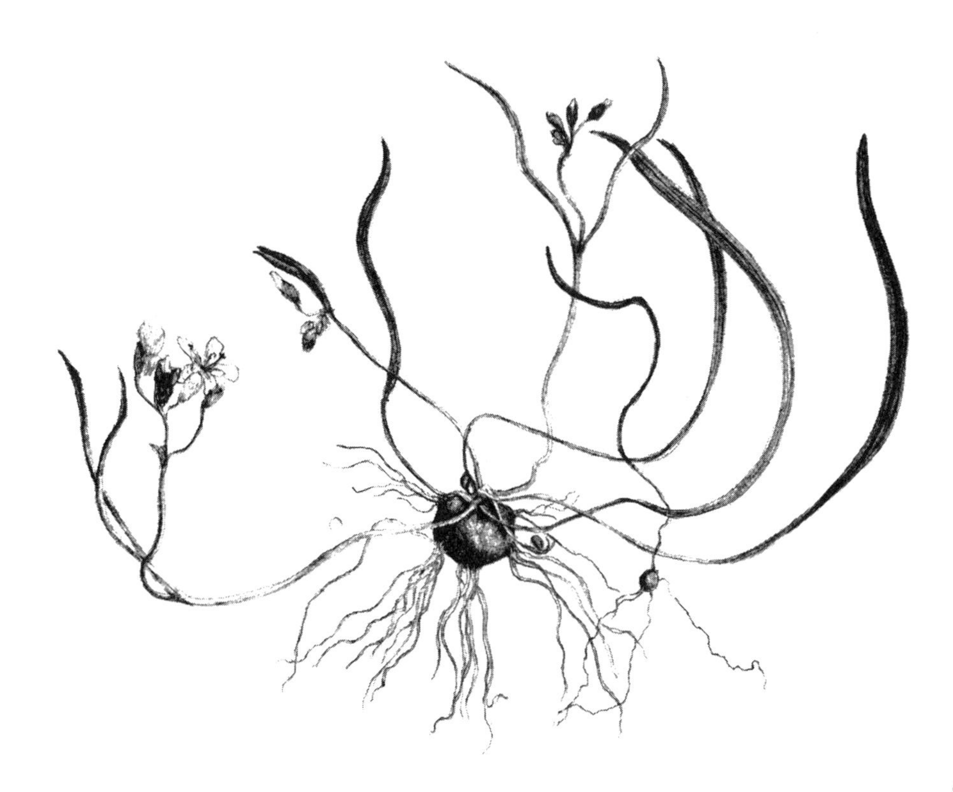

Allium tricoccum

Liliaceae, Lily family

OTHER COMMON NAMES:
Wild Leek, Three-seeded Leek

RAMPS

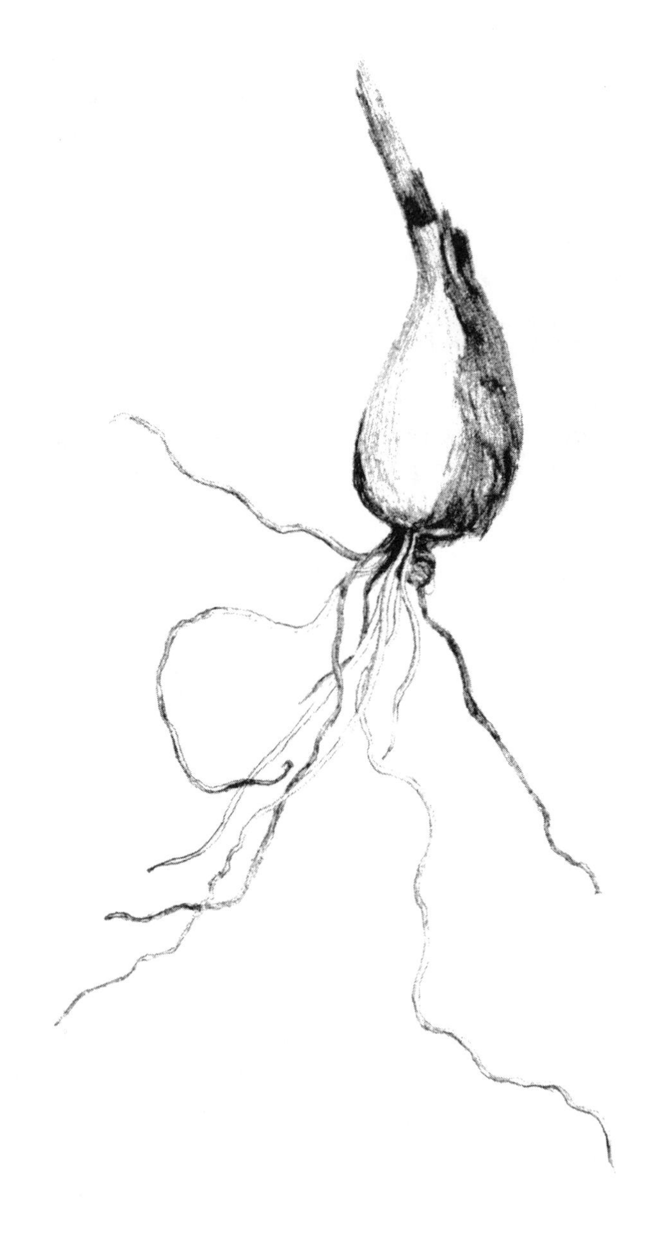

Ramps are found in rich, wooded areas from southeastern Canada to Georgia. Their distribution is usually limited, but where they are found they are often abundant.

The ramp is distinctive in that it first shows its two or three oblong leaves early in the spring. Usually by late March or April these leaves are unfurling out of the ground, but they wither and fall away before flowering time in June and July. The flowers are yellowish or cream colored and are arranged in a ball at the end of the single leafless one-foot long stem. Late in the fall round, shiny black seeds develop in capsules.

The name Ramps is another instance of an early English dialect word that survived in the semi-isolated communities of the Appalachian mountains. The word comes from Europe where it refers to the North American Ramp's English cousin, the ramson (*A. ursinum*) also known as the bear leek or wood garlic. The Swedes are credited with having named the plant ramson, or "Son of the Ram," because the Ramps usually appear under the sign of Aries, the ram, from March 20 to April 20.

Because Ramps are one of the first vegetables to appear in spring, they have been especially valued as a spring food and tonic by mountaineers, country people, and Indians ever since this continent has been inhabited.

Because of their strong odor, Ramps were called *pikwute sikakushia* (the skunk) in the language of the midwestern Menomini Indians. Rich woodlands near the southern end of Lake Michigan were said to be a favorite Ramp gathering area, for which reason they called this area *shikako*, (the skunk place). This is where the fuming, seething, metropolis, now known as Chicago, gets its name.

Wherever Ramps abound, their consumption is still a favorite springtime activity. Ramp-eating festivals sponsored by local granges, country churches, and community organizations serve as popular rites of spring in many rural areas. At these festivals there is fanfare along with multitudes of tongue-in-cheek references to the strong, onion-like odor of the Ramps. There is usually exaggerated window-opening in the festival hall, where boys who stuff themselves with as many Ramps as possible hope to be proclaimed too odiferous for school the next day. The Ramps are usually served fried, with traditional country fare of ham, beans, cornbread and greens. Ramps are also delicious any way you would use onions in soups, stews, or with sauteed vegetables. I don't think there could be anything better than a Ramp or two gently sauteed in butter with a few spring morel mushrooms (*Morchella esculenta*). Ramp omlette is another favorite of mine. The tender leaves also make a good addition to salads.

There is a popular belief that Ramps are good to eat only in the early spring because they get too strong later in the year. This is not true; in fact, since the Ramp season is so short, most of the Ramps I eat are "off season". When they are cooked, I don't consider them to be much stronger than a regular onion.

The bulb of the Ramp is usually very near the surface and often occurs in clumps of three or four together. The bulb in the illustration was dug in midsummer, when only the flowering stalk was above ground. The tunic wrapped around the bulb is probably the remains of the leaves.

Ramps give an excellent example of the bulb-type rootstock. The little stub under the bulbous part is actually the stem. The "rootlets" radiating out from it are the true roots, and the fleshy bulbous part is actually a cluster of thick leaves. A Cherokee method of harvesting Ramps is to cut or break off and replant this stubby stem as you gather the Ramps, so that they will grow again. Not only does this help conserve a valuable resource, but it also saves preparation time when you get ready to eat them, since you have to cut off this stem and the roots anyway.

Lilium superbum

Liliaceae, Lily family

TURK'S-CAP LILY

The Turk's-cap Lily, found in woods throughout the eastern half of the United States, is one of the most regal of our midsummer wildflowers. The stem is often seven feet tall and topped with several large nodding blossoms, each made up of six orange petals, six gracefully protruding stamens, and a pistil. The oblong leaves are arranged in whorls around the stem. The bulbs are in the form of thick fleshy scales arranged on a horizontal rootstock. At the base of the stem above the bulb are additional adventitious roots providing more support.

As food, this fleshy bulb and rootstock can be roasted or included in soups to good advantage.

There are several species of lily that resemble the Turk's-cap, and their bulbs can all be used similarly. None grow in abundance, so they should be gathered with care and moderation.

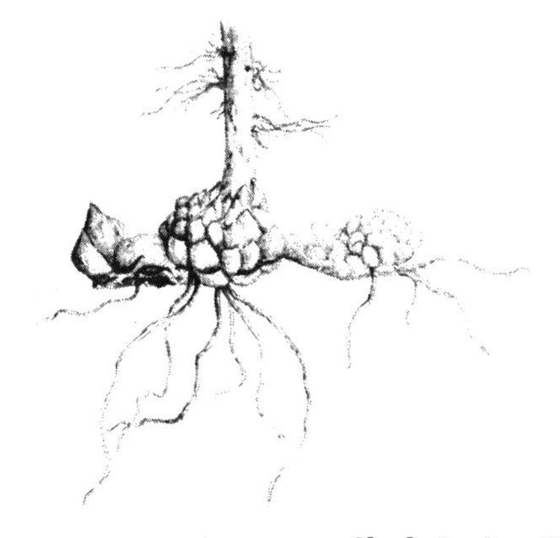

Aralia nudicaulis

Araliaceae, Ginseng family

OTHER COMMON NAMES:
Wild Sarsaparilla, False Sarsaparilla
Virginia Sarsaparilla, Shot-bush,
Small Spikenard, Rabbit's Root

AMERICAN SARSAPARILLA

American Sarsaparilla is found in poor to moderate soils, in woodlands, from Canada south to Georgia. This distinctive perennial grows about a foot high and has only one compound leaf which is divided into three parts. Each part usually bears five oval leaves. The straight flowering stalk is about eight inches tall and attaches underground at the base of the leafstalk. It usually bears three clusters of small greenish-white flowers in May or June. These are followed by round purplish-black berries having the appearance of bird shot, which probably accounts for the name shot-bush.

The name sarsaparilla, because it applies to so many different herbs and preparations, is a constant source of confusion. The sarsaparilla most generally found in commerce is the root of one of at least four different species of *Smilax* vine (the same genus as our green briars) native to various parts of Mexico, the West Indies, Central and South America. In fact, the name sarsaparilla comes from the Spanish *sarza*, meaning thorny, and *parilla*, a vine.

Our native American Sarsaparilla apparently acquired the same name because it has some of the same properties as the tropical briar. The trailing rootstocks are long and slender, often in the form of runners sending up several individual plants. They are somewhat aromatic, and from time to time I enjoy chewing a piece of it fresh out of the ground. The American Indians, some reports state, could subsist on these roots while on expeditions. I don't envy this aspect of their diet.

The infusion of the root makes a fair-tasting and quite warming tea. The rootstocks, like those of their tropical namesakes, have been used as ingredients in various root beers and tonics.

Medicinally, American Sarsaparilla is used for its tonic properties, and to treat fevers, chills and afflictions of the throat and chest. It was official in the United States Pharmacopea from 1820 to 1882 where it was classed as a stimulant, alterative and diaphoretic.

Cypripedium calceolus

Orchidaceae, Orchid family

OTHER COMMON NAMES:
Whipporwill's Shoe, Moccasin Flower, Noah's Ark, Venus Shoe, Venus Cup, Yellow Indian Shoe, American Valerian, Nerve Root

LADY'S SLIPPER

There are many varieties of *Cypripedium*, and most of them have similar medicinal properties. However, it is the Yellow Lady's Slipper *(Cypripedium calceolus)* that is the most commonly used as a botanical drug.

The Large Yellow Lady's Slipper is found in dry to moist woodlands and thickets east of the Great Plains, from Maine south to Alabama.

The plant usually consists of three to five, stemless, elliptical leaves with more or less parallel veins curving to a point at the top of the leaf. The leaves attain a length of six inches and are up to three inches wide. The whole plant is usually between one and two feet tall, topped with a uniquely beautiful yellow flower, with bulbous lower lip and spit curl sepals.

The blossom is certainly one of the most showy and beautiful of our native orchids. As the diversity of common names indicates, its shape has captured the imagination. The most commonly used name, Lady's Slipper, came from Europe as "Our Lady's Slipper," where it referred to a closely related woodland orchid which to this day in the French countryside is called "sabot de la Vierge."

The root is the medicinal part of the Lady's Slipper. As the names nerve root and American valerian reveal, this root has been highly regarded as a nerve tonic and sedative, used in a similar way as valerian root *(Valeriana officinalis)*.

It was official in the United States Pharmacopea from 1863 to 1916, and in the National Formulary from 1916 to 1936. It is classed as an antispasmodic, antiperiodic and diaphoretic.

An infusion of the rootstock is a standard method of using Lady's Slipper. It acts to tone and restore an exhausted nervous system and has been used to treat nervous depression, neuralgia, hysteria, epilepsy, insomnia, intermittent fevers and alcoholic delirium tremens. The American Indians used the herb similarly, and also to treat menstrual irregularities and to ease pains of childbirth. It can be combined with other herbs for more specific action. In combination with chamomile *(Anthemis nobilis* or *Matricaria chamomilla)* or other carminative-stomachic herbs, it is useful to settle a nervous stomach or nervous unrest due to stomach problems. For insomnia, it can be combined with passion flower herb *(Passiflora incarnata)* or hops *(Humulus lupulus)*. To help ease a nervous headache, catnip *(Nepeta cataria)* and/or scullcap *(Scutellaria lateriflora)* can be added.

The appearance of the ganglion-like, twisting roots radiating from the central perennial rhizome makes it seem that this plant might be appropriately named nerve root even if it wasn't useful as a nerve tonic. Perhaps this is one case in which we can apply the old Doctrine of Signatures which claims that somewhere in the physical appearance of the plant lies a clue, or "signature," as to its medicinal use.

To obtain maximum medicinal strength, Lady's Slipper root should be gathered in the fall and dried in the shade. It is far from being a common plant through most of its range. Protected in some states, it should be gathered very selectively, only in areas where locally abundant.

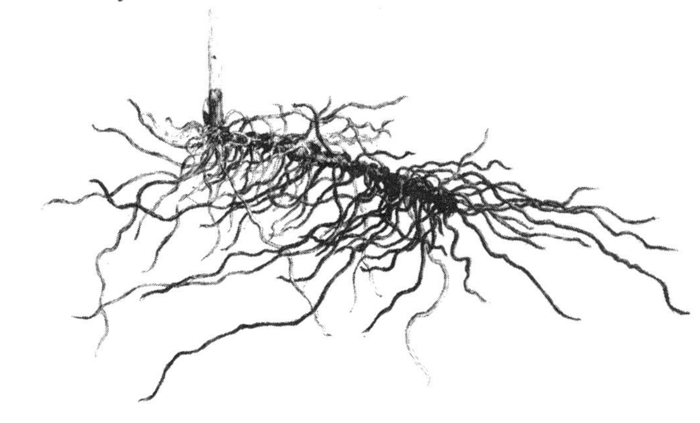

Aralia racemosa

Araliaceae, Ginseng family

OTHER COMMON NAMES:
Spignet, Petty Morel, Life-of-man,
Old Man's Root

SPIKENARD

Spikenard is found in moist woods throughout most of the temperate regions of North America, east of the Rocky Mountains. It is a handsome plant with a thick, arching stem three to seven feet tall and spreading branches with compound leaves of many leaflets. The large flower cluster (panicle) at the top of the plant bears dozens of tiny, white-petaled flowers in mid or late summer, and purplish-black berries early in autumn.

The thick, aromatic rhizome was official in the National Formulary from 1916 to 1965, where it was listed as a stimulant and diaphoretic. The decoction has been used as a blood purifier to treat all symptoms of impure blood, from syphilitic conditions to skin eruptions and acne.

Contained within the larger rhizomes are pockets of starch. When chopping up the autumn-dug roots I have nibbled on these and found them to be quite tender and mild flavored. This could make Spikenard a valuable plant as an emergency food.

Spikenard root can be made into an excellent cough syrup. Honey is used as the syrup base, and the Spikenard decoction or tincture is added either alone or in combination with other cough and sore throat herbs, such as wild black cherry bark (*Prunus serotina*), elecampane (*Inula helenium*), coltsfoot (*Tussilago farfara*). It has also been used to ease the pains of childbirth, taken as a tea in the months before delivery.

The rootstock is thick and very extensive. Usually it is necessary to dig only one medium size plant to fill the needs of several people for quite sometime. The pleasant aroma of the fresh rhizomes make up for the effort that it sometimes takes to unearth one of the plants.

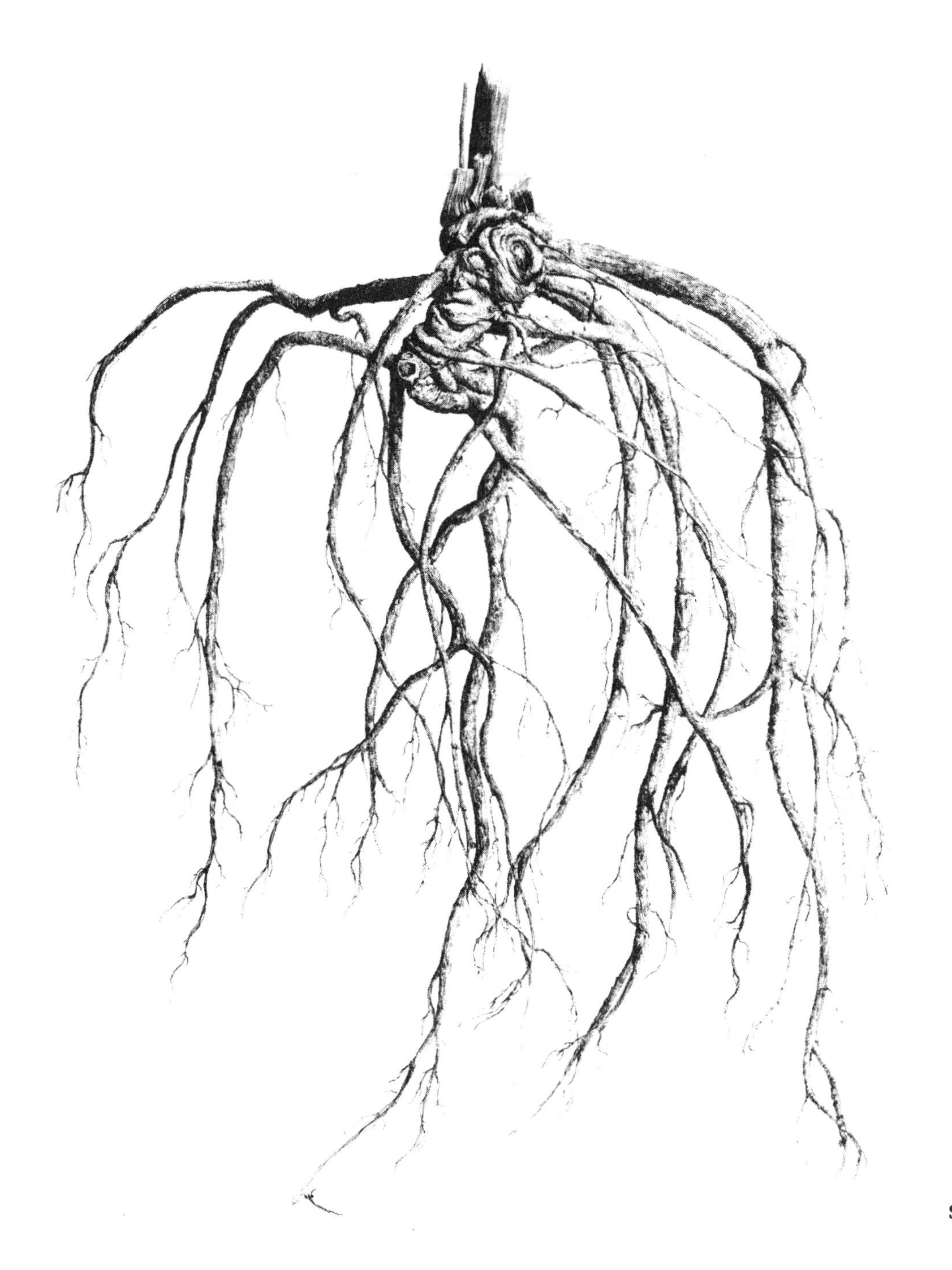

Adiantum pedatum

Polypodiaceae, Fern family

OTHER COMMON NAMES:
Maiden Fern, Rock Fern, Hair Fern

MAIDENHAIR FERN

Maidenhair Fern is found throughout most of the eastern half of the United States in moist, shady and rich woods, especially in limestone areas.

The Maidenhair is a graceful and delicate fern, growing twelve to eighteen inches tall with small, alternately arranged leaflets along a distinctively branched stalk, polished black to deep reddish brown. It is probably the gracefully flowing shape and textural qualities of the non-fern-like shiny stem that gives the plant the name Maidenhair Fern.

The rootstock of this plant has been used by both the Indians and white men in an infusion or decoction as a bitter tonic, as a stimulant or expectorant to treat respiratory ailments and to soothe throat inflamations. The green parts of the plant are used similarly.

The perennial rhizome, typical of most fern rhizomes, is covered with scales which are actually a form of underground leaf. This helps us remember that a rhizome is in fact a stem.

Other types of ferns have useful rhizomes. The most notable are various species of the genus *Dryopteris*, called male fern or aspidium. Male fern has been used to expell worms, especially tapeworms and is taken as a fluid extract or as an oleoresin. It was listed in the United States Pharmacopea until 1965, but has come into disuse because in too large of a dose, it is an irritant and poison, and therefore not recommended for uninformed self-medication.

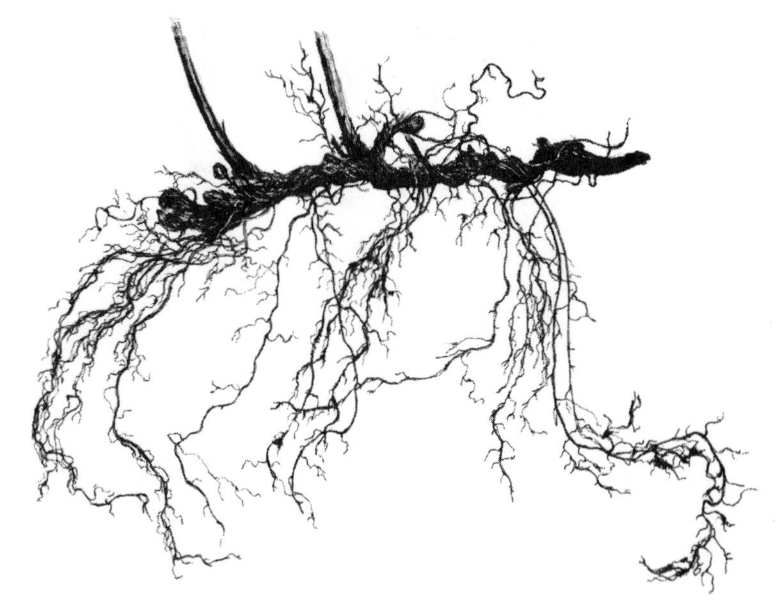

Gillenia trifoliata

Rosaceae, Rose family

OTHER COMMON NAMES:
Bowman's Root, American Ipecacuànha, False Ipecac, Western Dropwort, Indian Hippo

INDIAN PHYSIC

Indian Physic is found in woodlands with poor to moderate soils in hilly or mountainous areas, from southern Canada and Michigan to New York, and south to Georgia and Alabama.

The slender, reddish stems of the plant are about two or three feet tall; the compound leaves are almost stemless and are in groups of three toothed leaflets. They somewhat resemble blackberry or dewberry leaves (*Rubus spp.*) in texture and structure. From May to July, the nodding flowers are produced in loose, terminal clusters and have five white or pinkish petals. The seeds are formed in small, five-divided capsules.

The roots of Indian Physic were official in the United States Pharmacopea from 1820 to 1882, where they were listed for their emetic properties.

As with many other botanical drugs, white men learned to use this plant from the Indians. This is where it got such names as Indian Physic, and bowman's root. It is said to resemble, in action, the Brazilian root *Cephaelis ipecacuanha*, known as ipecac.

In small doses the powdered root is tonic, but in large doses it induces vomiting in a gentle manner, if that is possible. It has been used to treat dysentery, stomach problems, rheumatism, worms and fevers.

The perennial rhizome often produces several flowering stalks. The roots are spreading and shallow. They should be gathered in the fall.

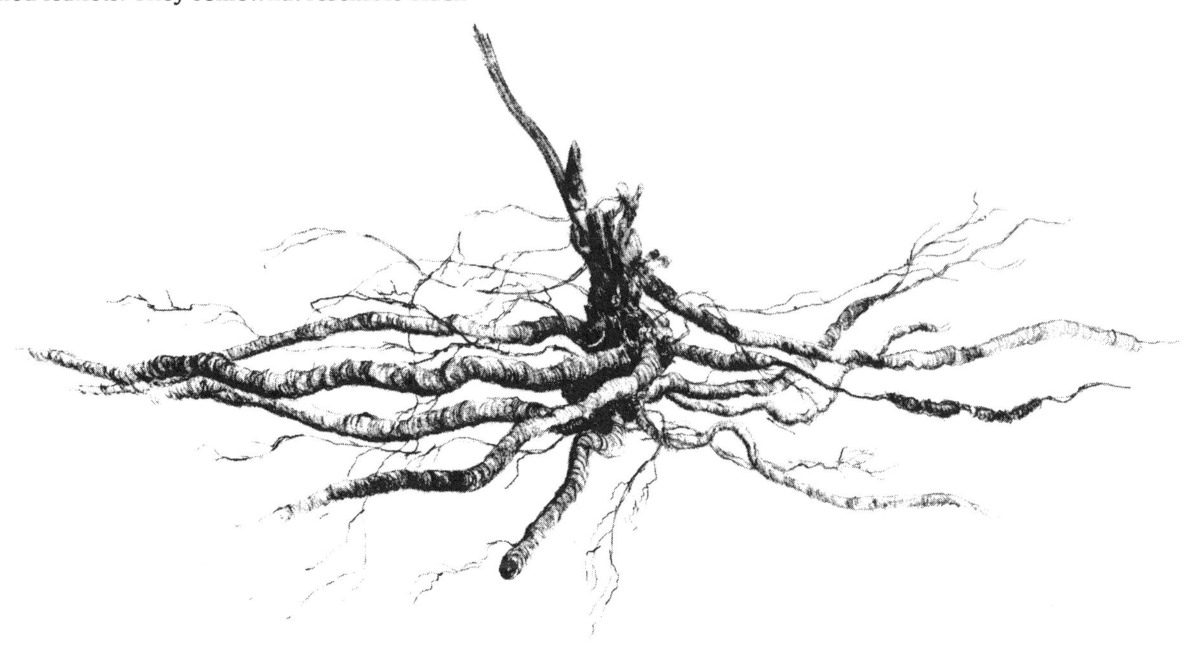

Asarum canadense

Aristolochiaceae, Birthwort family

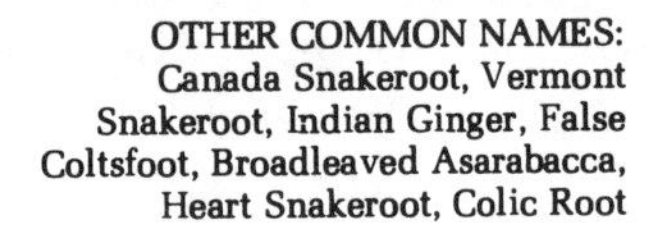

OTHER COMMON NAMES:
Canada Snakeroot, Vermont Snakeroot, Indian Ginger, False Coltsfoot, Broadleaved Asarabacca, Heart Snakeroot, Colic Root

WILD GINGER

Wild Ginger is found in moist, shady forests, especially in limestone areas from southeastern Canada southward into the mountain regions to Georgia.

The soft, downy, heart-shaped leaves of the Wild Ginger can be seen, often near streams, in beds or colonies. The leaves occur in pairs, each leaf four to seven inches across. They rise directly from the rhizome to attain a height rarely more than a foot above ground.

In April or May, at the crotch of the two leaf stems, the keen observer may find a unique but inconspicuous flower. Composed of three, petal-like lobes that are rich reddish-brown or maroon in color, the flower dangles from a short, flaccid stem that can barely keep it off the ground. Sometimes, in fact, the flower blooms while buried under old leaves and must be uncovered to be seen. The fruit which follows is a fleshy six-celled capsule.

Though I can find no reference indicating that Wild Ginger has ever been considered useful as a remedy for snakebite, I feel it came to be called snakeroot because other related plants, in this country and elsewhere, do have this reputation. The name false coltsfoot is derived from the hoof shape of the leaf, and of course, its resemblance to true coltsfoot (*Tussilago farfara*). The name asarabacca comes from England. It is the name of the American Wild Ginger's similar European cousin (*Asarum europaeum*).

It is truely a blessing that in our cool, northern latitude forests we have a plant with the same taste and aroma, and so many of the same properties, as the exotic tropical spice. Botanically, however, the two gingers are not closely related.

The rhizomes of the Wild Ginger can be used in several ways. They can be candied like the ginger of commerce by first boiling the pieces of the rootstock in water, until tender, and then adding sugar or honey and cooking a while longer (about 30 minutes) and draining. You will then have Wild Ginger honey or syrup, and pieces of candied Wild Ginger. The pieces of ginger can be stored indefinitely in airtight jars. Some people like to roll the pieces of ginger in white granulated sugar to make them fancier, though less healthful. The syrup can be used to sweeten drinks; to pour over ice cream, yogurt, or pancakes; as an ingredient in sweet-sour sauce; or wherever else you use syrup. The candied roots make a spicey nibble for almost any occasion. But since they are carminative, meaning that they settle the stomach, they are particularly appropriate after a heavy meal of foods that tend to form gas. The rootstocks can be ground to a powder and used like commercial ginger with apple dishes or to spice up sauteed vegetables. On more than one occasion, I have sat euphorically contemplating the "good life" with my mouth full of maple-ginger bread—homemade with Wild Ginger and maple syrup.

Medicinally, the roots and rhizome of Wild Ginger are classed as an aromatic stimulant, carminative, tonic and diaphoretic. It was listed in the United States Pharmacopea from 1820 to 1873, and in the National Formulary from 1916 to 1947. It is used in cases of flatulence, digestive upset, spasms of the bowels, and colic, hence the name colic root. Its diaphoretic properties make it useful in colds and fever. The American Indians have used Wild Ginger in all of these ways, as well as an emmenagogue and parturient for women.

It is reported that two active antibiotic substances have been isolated from the plant.

The rootstocks of the Wild Ginger, because they grow in loose, rich soil very close to the surface of the ground, are one of the easiest roots to dig. And their wonderful, spicey aroma, especially when combined with the rich earthy smell of the soil, make them one of the most pleasurable roots to gather.

Dioscorea villosa

Dioscoreaceae, Yam family

OTHER COMMON NAMES:
Colic Root, Rheumatism Root, Devil's Bones

WILD YAM

This graceful, twining vine is found in moist woods and thickets east of the Great Plains, from southern New England south to Florida and Texas.

The vine springs forth new every year from the perennial rootstock to attain a length of up to fifteen feet. The heart-shaped leaves are borne on long leaf stems and are either alternate, opposite, in whorls, or a combination of these arrangements. The drooping clusters of tiny greenish-yellow flowers are in bloom about mid-summer.

The fruit appears as three-winged capsules containing flat, dark brown, winged seeds that ripen in September and remain on the vine well into the winter. They add a unique element to dried flower arrangements.

The Yam family is a very important group of plants. Some 600 species are known. Various Yam species are staple foods in many parts of the world, especially the tropics.

Certain other species, most notably the Mexican Wild Yam, have been shown to be a source not only of testosterone and cortizone, but also of diosgenin, the basic material from which some birth control pills and other steroid drugs are made.

Our native, northern latitude Wild Yam has not earned this fame, but it was official in the National Formulary from 1916 to 1942, classed as a diaphoretic and expectorant.

A decoction of the rhizome is recommended against all forms of colic, especially bilious colic, hepatic congestion and other forms of intestinal and abdominal irritation. It is considered an anti-spasmodic, useful in treating spasms of all kinds, including spasmodic asthma and vomiting. In small frequent doses, it is known to relieve the nauseous "morning sickness" characteristic of early pregnancy and to help ease the entire prenatal time. For this, it is recommended to combine the Wild Yam with other "squaw tonics" such as raspberry leaves *(Rubus ideaus)* and squaw vine *(Michella repens)*.

Wild Yam has also been used to treat rheumatism.

In large doses, it is considered diuretic, but can also be emetic.

The large, hard rhizome has numerous stem scars along its length and many characteristic wirey roots extending from it. The shape and hard, bony texture of the dried root probably accounts for the name devil's bones.

The rootstock in the drawing on the title page of this book was almost two feet from end to end, and you can see that it had attained sufficient strength and size to send up two stems.

The Chinese yam *Dioscorea batatas*, also known as Cinnamon vine, was originally brought here from the Orient and planted as an ornamental vine. It has since become naturalized and can be found along back roads and around old home sites throughout most of the southeastern portion of the country.

It has strings of tiny, white, cinnamon-scented flowers that bloom in midsummer. The leaves are usually halberd or fiddle-shaped. The plant rarely produces seeds but late in the year it forms tubers, not underground like most plants but in the axils of the leaves, often well above the ground. These tubers resemble miniature potatoes, about the size of marbles. They can be gathered easily and are delicious when gently cooked by rolling them around on a hot, well-oiled frying pan.

Deep in the earth the vine is rooted by a pair of large tubers. One is usually fleshy and succulent, the other is somewhat dried and shriveled. Presumably this represents successive years growth. The fleshier tuber is delicious cooked any way you would cook potatoes. The ones I have gathered have been about the size of small to medium potatoes but there are reports of them weighing several pounds. (See illustration opposite.)

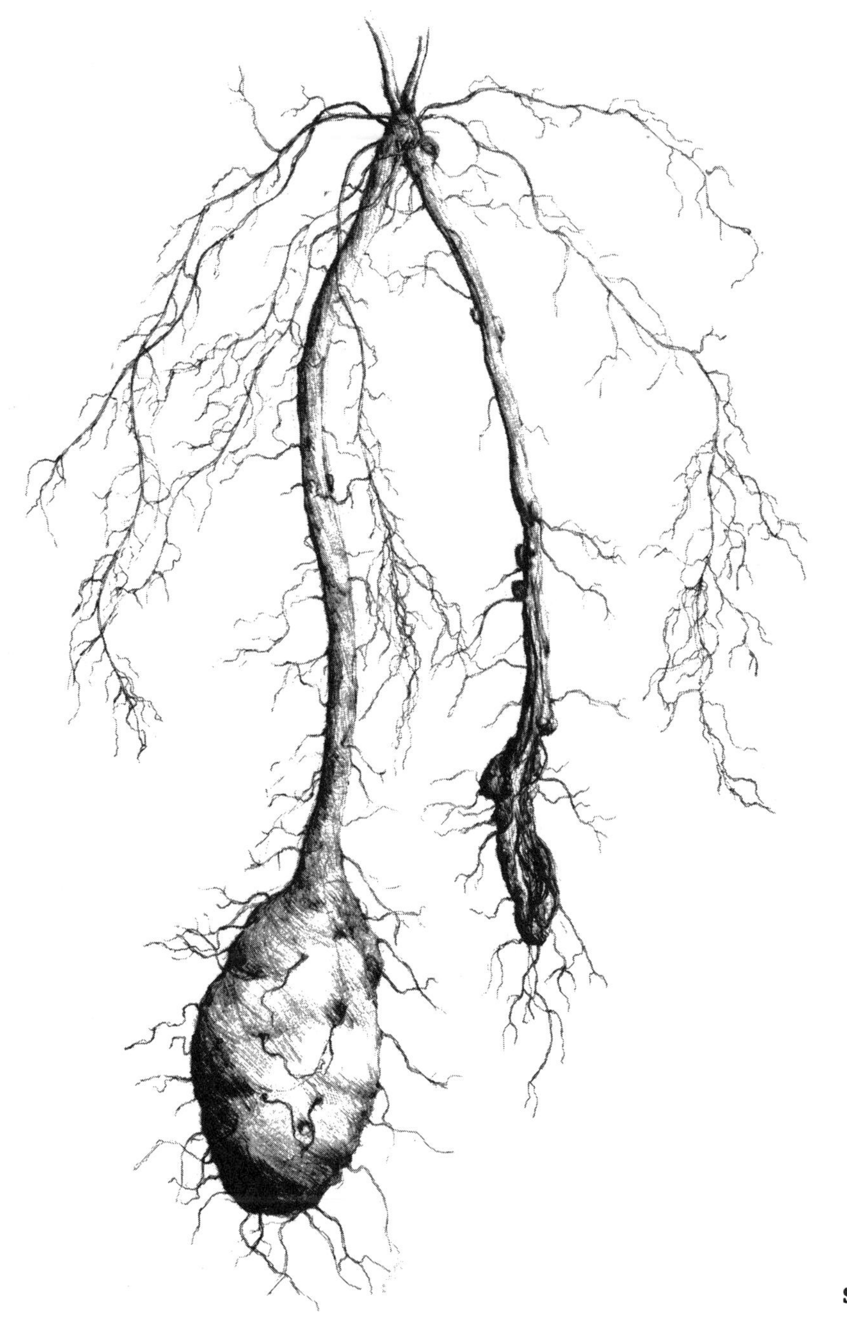

Corallorhiza odontorhiza

Orchidaceae, Orchid family

OTHER COMMON NAMES:
Crawley Root, Dragon's Claw, Chicken Toe, Turkey Claw, Fever Root

CORAL ROOT

Coral Root prefers fairly dry shady woods and is usually found growing in rich deposits of leaf mold throughout the eastern part of the United States.

This small member of the orchid family is usually less than a foot tall and manifests itself as a stalk topped with a raceme of six to twenty tiny, purplish flowers blooming between July and September. These are followed by ribbed, somewhat egg-shaped seed capsules. The leaves are reduced to sheath-like scales on the stem and are colored purplish-brown to almost magenta. There is no green coloration in any members of the genus *Corallorhiza*. Chlorophyll is almost totally lacking in these plants, which means they do not manufacture their own food and almost totally depend on a fungus which infects their roots to bring in nutrients from the surrounding leaf mold.

The peculiar nature and jointed, coral-like shape of the root gives the plant most of its common names. The name crawley root, however, is probably a corruption of coral root, coming from the fact that the average woodsman knew nothing of sea coral.

The name fever root, comes from its use in herbal medicine. It is acclaimed as one of the most sure and fast-acting diaphoretics, promoting free perspiration without producing other heat or excitement in the system, and has found much use in treating diseases with which fevers are associated. In addition to being diaphoretic, it also is a gentle sedative. These properties are particularly useful in combination with other herbs aimed at treating specific problems.

Coral Root has been combined with the resin of blue cohosh root and used to regulate faulty menstruation and to relieve pain and febrile conditions associated with childbirth.

Combined with May apple resin, or extract of Culver's root, it has been used as a remedy for liver and bowel problems. To treat flatulent or bilious colic, it is combined with wild yam. There are several species of Coral Root, and it is believed that all have similar properties. The fresh root has a peculiar, strong smell and white color. When the root is dried, its odor dissipates, and its wrinkled, dark brown pieces compare to cloves in appearance. Because of its small size and high price, Coral Root is rarely sold these days. If you do gather your own, it is recommended that it be pulverized, well-dried, and stored in an airtight container away from light.

Aplectrum hyemale

Orchidaceae, Orchid family

OTHER COMMON NAMES:
Putty Root

ADAM AND EVE ROOT

This small plant is found in the moist, heavy soil of shady woods through most of the East, from New England to Georgia.

This is one of the most interesting of our low-key native orchids. In the spring the single flower stem arises, bearing up to fifteen small, pale, greenish or yellowish blossoms variably marked with purple and having white lips. The entire flowering stalk rarely exceeds fifteen inches in height. As autumn approaches, the flowers form into seed capsules. As these capsules ripen, they dry and disperse their multitudes of tiny seeds. Meanwhile, out of the ground and a few inches from the stalk appears a solitary, elliptical leaf. This leaf has thin, white pinstripes, is folded like a pleated skirt, green on the top side, with a tinge of purple underneath. There that leaf remains until spring, when it dies as the new flower stalk arises at it's side.

A look underground will give more insight to what is going on. The flowers emerge from a fleshy, glutinous corm. As the plant flowers above ground, that underground corm is forming another corm connected by a short underground stem (stolon), which will produce the leaf in the autumn and the next year's flowering stalk.

Because the two corms are so neatly linked together, the plant came to be called Adam and Eve Root. However, as with human relationships, this plant's relationships and growth patterns cannot be accounted for in such a simple manner. In the particular autumn-dug root I have illustrated, it would appear that Adam (or Eve, as you like) is carrying on a bigamous relationship with two spouses! And who is that shriveled old babe in the background? If we can turn from our anthropomorphizing and back to botanizing, it will come clear. The fact that there are two new autumn leaf corms is an indication that the plant is in good health, good soil, or is having a good year or most probably a combination of the three. The wrinkled corm in the background is from the previous year. It has already produced a leaf and flower stalk that have died and returned to the soil, and the corm itself is in the process of decay.

The Adam and Eve Root got the name putty-root because the glutinous texture of the corm makes it useful as a cement or putty. It was used as such by early settlers who were fortunate enough to have glass for their cabin windows.

The plant has had very limited medicinal use. It has been used to treat bronchial illnesses and could be a soothing herbal base for making poultices.

In occult lore, Adam and Eve Root has a cupid-like reputation for helping to magically maintain the bond between lovers. In a solemn ceremony, the two lovers each receive one of the corms. The woman takes into her possession the male, or Adam root, and the man takes the female, or Eve root. The "sex" of the root can be told either by the shapes of the roots, or by the fact that the female root-corm bears the flower, and the male the leaf. As the legend goes, so long as the pair maintains possession of their respective roots, their bond shall remain strong and true.

In these times of raging promiscuity, soaring divorce rates and tumultuous interpersonal relationships, we can't afford to snicker at a potential herbal cure for these maladies. I have been experimenting with this use of Adam and Eve Root for quite some time now, and all four of the roots in my possession have been working well .

ROOTS OF SUN-LOVING FIELD PLANTS

A substantial number of the plants I have covered in this group are not indigenous to North America. The fact that they frequent sunny areas in the disturbed soil of fields and roadsides is not surprising. In the woodlands the complex variety of vegetation is made up of species which have evolved together. Through natural selection they have been fitted to each other's company like pieces of a jigsaw puzzle. When any of these plants go to seed the chances are good that the seedling will find an appropriate ecological niche, and grow to maturity. If an alien seed from some other region or flora is planted there, even though it might be able to germinate, its chances of survival are extremely slim simply because it does not fit into the tightly interlocked system of that vegetation. This is a closed habitat.

Disturbed soils, on the other hand, are an open habitat. There is less interference from the native plants. Cultivated fields, railroad right of ways, roadsides, dump heaps or cattle barns are the cause of the disturbance, the result is a change in the quality and character of the soil. This creates an open habitat which allows the aliens to move right in and get established. These open habitats are, of course, closely associated with human settlements. In most cases it has been the settlers who brought the seeds to the new habitat in the first place — either purposely, as ornamental or medicinal plants, or accidentally from seeds in pant cuffs, stuck to clothes, or mixed in food or fodder.

Inula helenium

Compositae, Composite family

OTHER COMMON NAMES:
Elf Dock, Elfwort, Horseheal, Scabwort, Horse Elder, Yellow Starwort

ELECAMPANE

Elecampane was introduced here from Europe, to be planted and naturalized in the colonists' medicinal gardens. In succeeding years it has taken to wandering about on its own and can be found growing in fields and along roadsides from eastern Canada south to the Carolinas and Missouri.

The flowering stalk of the Elecampane is two to six feet tall and rises out of a basal clump of broadly oblong fuzzy leaves that are ten to 20 inches long. The flowers are in heads looking somewhat like a stout, yellow, sparsely petaled daisy or sunflower, two to four inches in diameter.

Elecampane has been known and used since antiquity. Pliny, in the first century A.D., advised Augustus and Julia to "let no day pass without eating some of the roots of Enula, considered to help digestion, expel melancholy and sorrow, and cause mirth." That sounds good already.

I ate some of the candied roots made by friends and found the taste very aromatic and camphorous. I could see how they would be settling to the stomach and how the camphorous vapors would be helpful in clearing up a cold. Although I wouldn't say they were unpleasant, I found them about as mirth producing as eating Vick's-vapo-rub. The candy, made with more sweetener and less Elecampane, was a favorite of 19th century English schoolboys who believed it to give agility and endurance in athletics.

It was listed in the National Formulary and is considered a stimulant, diaphoretic, diuretic, expectorant, tonic and emmenagogue which makes it useful in treating respiratory disorders of all kinds, as well as menstrual irregularities. The plant also seems to have some antiseptic and bactericidal properties.

The name scabwort comes from the fact that a decoction of the root has been used to cure sheep of a disease called *scab*, and the name horseheal from its use in treating skin diseases of horses.

An old legend says that Elecampane was born of the tears of the beautiful Helen of Troy when she was stolen away by Paris. Wherever her tears fell the plant sprang forth. Its name is supposed to be a very old contraction of *Elena-de-la-campagna*, or Helen of the fields.

In Ireland, Denmark and other parts of Europe the plant acquired the name elf dock because, long ago, there was a very poor maiden who fell in love with a handsome prince who she saw hunting in the forest. But, alas, she was a mere peasant not worthy of a future king's attentions. As was the custom among her people, she put the nightly bowl of milk on the door step for the wee folk and she sat down beside it to pine and sigh. When the elves came for the milk they found her there sound asleep. Being both kindly and magical, they put a spell on her and wisked her off into the forest where they clothed her in the large coarse leaves of Elecampane and placed a chain of the yellow flowers around her neck. At dawn, when the prince came riding through the woods, her gown had been transformed into the richest green velvet and her necklace to the finest gold. The prince was awe struck by her beauty, but not so overwhelmed as to forget to woo her and carry her off into the sunrise to be his, happily ever after.

The moral of the story, I presume is: Never underestimate the power of a weed.

The large, thick, perennial root should be gathered in the first few years of its growth before it becomes too woody, and it should be dried carefully before it is stored, as it is prone to mold.

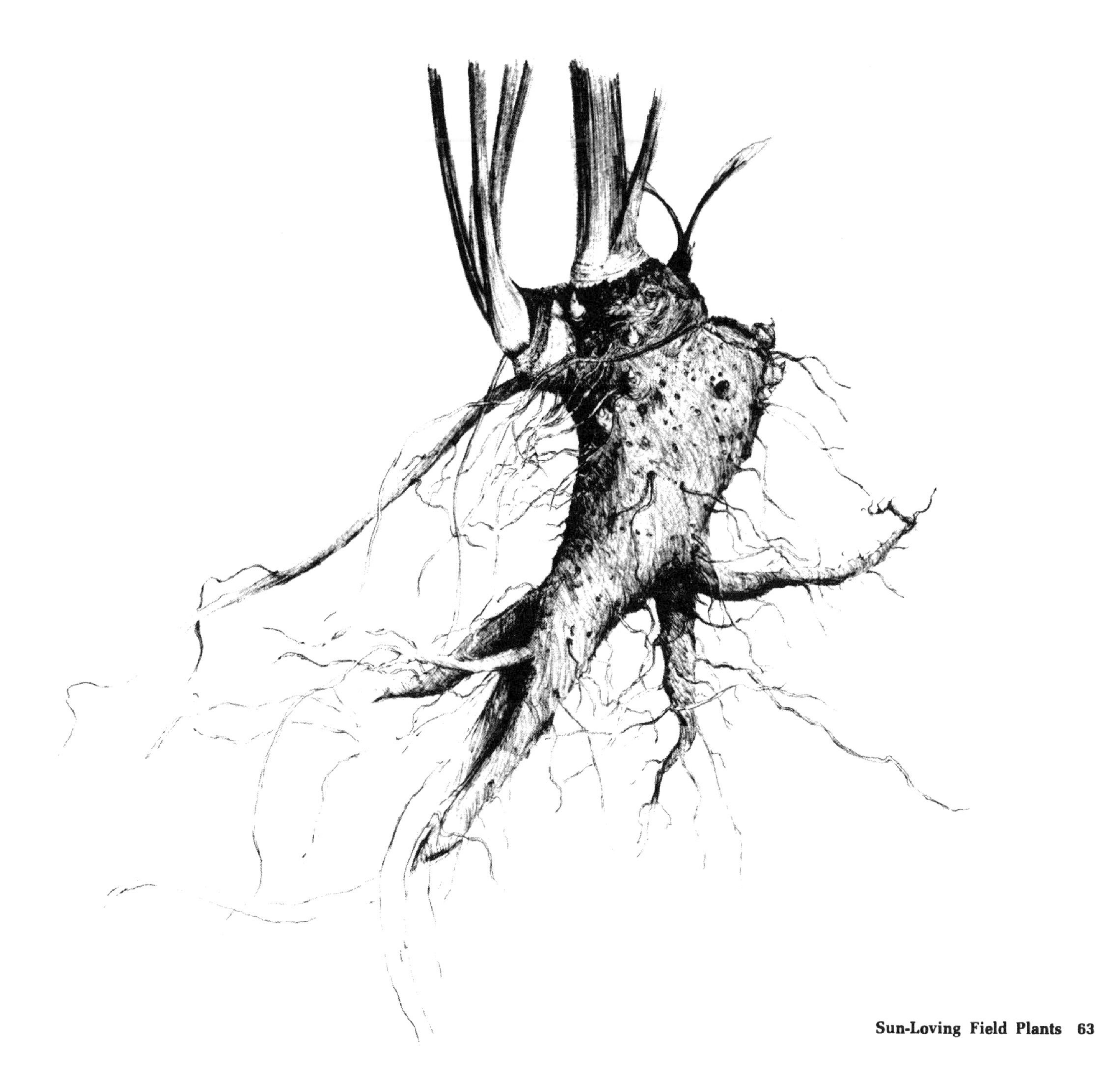

Gelsemium sempervirens

Loganiaceae, Logania family

OTHER COMMON NAMES:
Carolina Jasmine, Carolina Wild Woodbine, Evening Trumpet Flower, False Jasmine

YELLOW JESSAMINE

Yellow Jessamine is a tropical vine found in the edges of woods and thickets from Virginia south into Florida, west to Texas and down into Mexico and parts of Central America.

The height of this vine depends on its support. It can be found on the ground, climbing along fences, or up in the higher branches of trees.,

It has opposite lance-shaped evergreen leaves. The blooming season can occur anywhere between January and April. Wherever the bright yellow blossoms are abundant, their fragrance fills the air and is one of the finest announcements that spring has arrived. It is best, however, that only the fragrance and visual beauty of the Jessamine flowers be sampled, because both the flowers and the honey made from them is regarded as poisonous.

The rootstock, which is gathered after the plant has flowered, is the part considered to have medicinal use and was official in the United States Pharmacopea from 1863 to 1926, and in the National Formulary from 1926 to 1955.

It has been used primarily for its depressant effect on the central nervous system as an antispasmodic, sedative, nervine and also to dilate the pupils. Because an overdose of this drug can be fatal, it is recommended that its use be limited to experienced medical practitioners.

Yellow Jessamine is the South Carolina state flower.

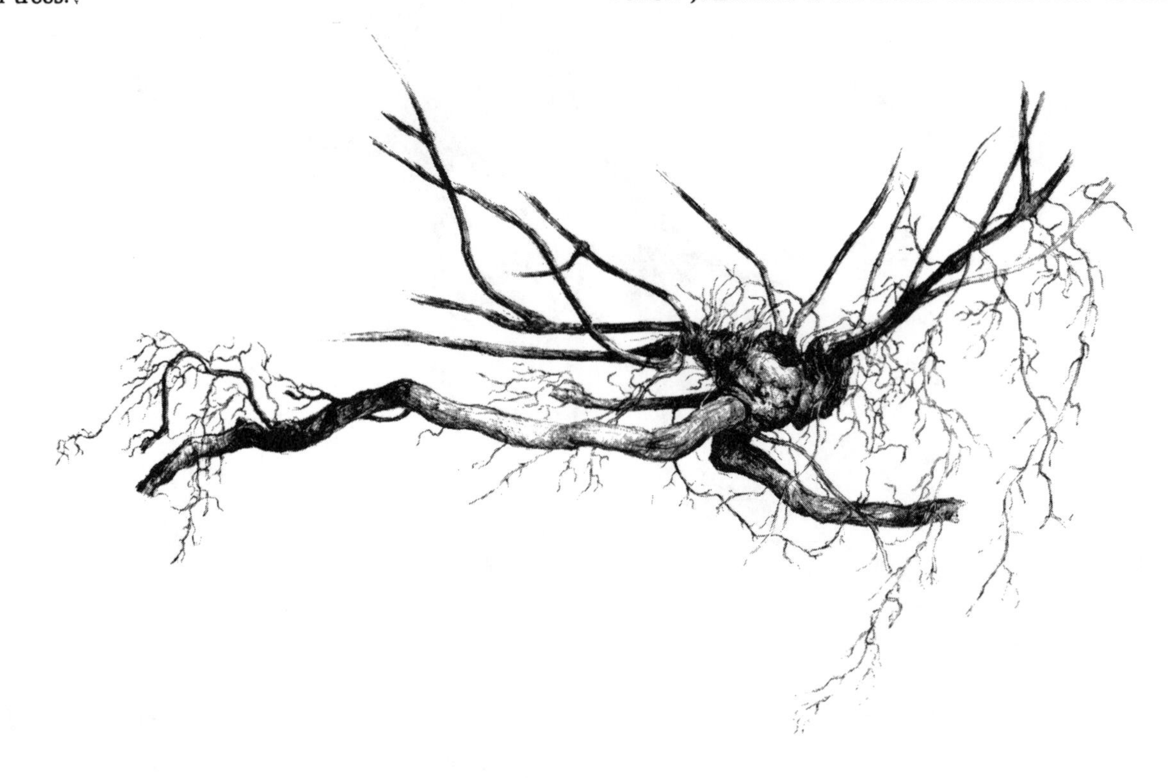

Aletris farinosa

Liliaceae, Lily family

OTHER COMMON NAMES:
Star Grass, True Unicorn Root, Ague Root, Rheumatism Root, Crow Corn, Mealy Starwort, Bitter Grass

COLIC ROOT

Colic Root is found from Maine to Florida and west to Arkansas in moist, sandy meadows and savannas, often near bogs and pine barrens. It is most abundant on the coastal plain in the southern portion of its range.

The flowering stalk of Colic Root is two or three feet tall, and except for a few small, bract-like leaves, it is bare. Hugging the ground is a proportionately small rosette of grasslike leaves, each two to six inches long, from which the plant rises. The erect flowering spike of flowers, from four to twelve inches long, is produced from May to July. These urn-shaped flowers are white, sometimes tinged with yellow. They appear to be covered with a fine mealiness resembling flour. The many-seeded capsules are ovoid and divided into threes.

Medicinally, the dried rootstock, in a decoction or tincture, is used as a valuable bitter tonic for the stomach to relieve flatulence, indigestion and colic, as the name, colic root indicates. It is highly valued for its beneficial influence on the female generative organs, but also as a general tonic and in cases of habitual miscarriage.

Aletris was official in the United States Pharmacopea from 1820 to 1873, and in the National Formulary from 1917 to 1947.

The perennial rootstock is short, rough and scaly and is almost completely hidden by the fibrous roots and the remains of the leaves. It is best gathered in autumn.

Pueraria lobata

Leguminosae, Pea family

KUDZU VINE

Kudzu (pronounced cud-zoo) Vine has been planted for erosion control, as a forage crop, and an ornamental vine throughout most of the southeastern states. Its rapid growth (40 to 60 feet per season) and its tenacity has been highly touted by soil conservation and extension agents. But like many agricultural quick cures, it is now the bane of many farmers and land owners because once established, it is incredibly fast spreading and almost impossible to eradicate. Kudzu is often planted along the scarred banks of new highways. After a few years, it can be seen spreading over huge areas, often completely covering trees and sometimes even barns and houses with its lush, tropical-looking growth. The leaves are compound, a foot or more long, with three lobes resembling oversized bean leaves. The flowers are purple and very sweet smelling. Perhaps to compensate for its vigorous activity in the summer, it is highly sensitive to freezing. The first touch of frost stops its growth and Kudzu retires for the season.

The key to its tenacity is its large root which usually divides into several, thick, three to five foot-long branches.

What interested me about these roots was a report which indicated that in Japan quantities of Kudzu roots are dug, "often with great difficulty," whereupon "they are cleaned, cut in pieces, crushed and the starch washed out and allowed to settle to the bottom of the tub. The starch is then purified by repeated washings and when dried is a fine, pure white article much esteemed for food." (7)

Following this sparse lead, I dug a Kudzu root. It had three swollen main branches about three feet long. It came out of the ground with a bit of difficulty, like the report said. After washing it well, I chose the fattest branch and chopped it into chunks with a hatchet. These chunks I put into a large stoneware bowl, covered them with water, and commenced pounding them with a wooden pestle. The water turned dark brown and the root chunks were very fibrous. Though I could see what I thought to be starch clinging to the fibers, separating it from the fibers was not so simple. I would pound for a while, and then would let it settle for a while. After it settled, I would carefully pour the water off—watching for a sediment to form on the bottom. After repeating this cycle for most of the day I started to notice (joy of joys) a little bit of white starch that was indeed settling out on the bottom. With this to give me second wind, I redoubled my efforts, going well into the night, pounding Kudzu.

The next morning, I removed all the pounded fibers and after it had settled well, I carefully poured off the water. This left me with about half of a cup of watery sediment. Not bad for a day's work? To evaporate the rest of the water, I poured it into a saucepan and placed it on the back of the wood stove. A little while later, presto! It had turned not into the fine white flour I had expected, but into a translucent, gelatin-like material similar to corn starch. Not knowing what else to do with it, I added a little milk and honey, to form a sort of soupy pudding which was pleasant tasting, but certainly not worth the work! I have postponed any further experiments with Kudzu till I acquire a water-buffalo-powered Kudzu crusher similar to what may be used in the Orient.

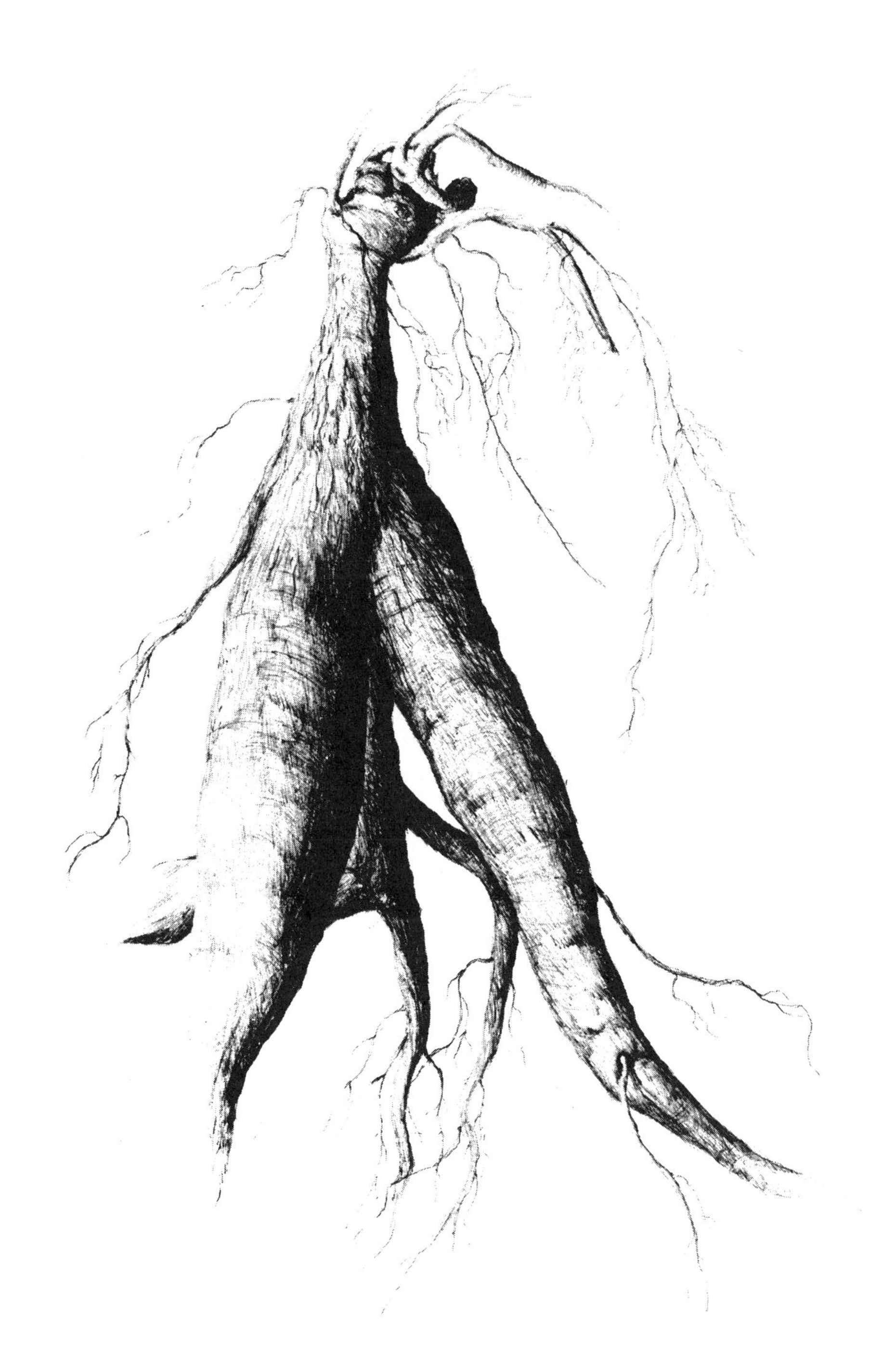

Apios americana

Leguminosae, Pea family

OTHER COMMON NAMES:
Indian-potato, Ground-potato,
Potato-pea, Pig-potato, White Apple,
Traveler's Delight, Wild Bean, Bog
Potato

GROUNDNUT

Groundnut is found at the edges of fields and thickets, usually in moist bottomland near streams. It ranges from Canada west to Minnesota and Colorado, south to Texas, Louisiana and Florida.

Groundnut is a vine which can be found spreading over the ground or climbing over bushes to a height of several feet. The compound leaves have five to nine leaflets. In late summer the blossoms spring forth in clusters on spike-like racemes from the junction of the leaf stalks and the stem. The flowers are brownish purple, very fragrant, and shaped like a bean blossom. In autumn the blossoms form slender, bean-like pods.

Groundnuts were a staple food of the American Indians. Both pilgrims and early colonists often subsisted on them.

The great botanist, Asa Gray, once said that if advanced civilization had started in North America instead of the Old World, the Groundnut would have been the first tuber to be developed and cultivated. Actually, they were taken to France to be cultivated as early as 1635 but were soon forgotten, probably because the roots take two or three years to reach eating size. In Captain John Smith's *Narratives of Early Virginia*, he tells of "Groundnuts as big as Egges and as good as Potatoes, and 40 on a string not two inches underground."

I have never found them quite like Smith described, but then again, he got there first.

Now and then the Groundnuts I find approach egg size (but not Grade A Large). I never found 40 on a string, but I have to agree that they are as good as potatoes—even better, I'd say. So Groundnuts haven't changed so much after all.

Groundnuts can be eaten raw or boiled, but I like them best sliced thin with the skin on and gently fried in a little oil with a bit of wild onion or garlic added for additional flavor. They are a relative of the soybean and are reported to contain 25 percent protein.

Perhaps a search for these delicious nuggets and other riches of our back-pasture treasure houses stir up in us the feelings of John Greenleaf Whittier when he wrote, in *The Barefoot Boy:*

Where the whitest lilies blow,
Where the freshest berries grow,
Where the groundnut trails its vine,
Where the wood-grape's clusters shine.

Taraxacum officinale

Compositae, Composite family

OTHER COMMON NAMES:
Priest's Crown, Swine's Snout, Lion's
Tooth, Blowball, Fortuneteller,
Doonhead-clock

DANDELION

Dandelions are so abundant and familiar that they hardly need description. Most of the names relate to its looks or phases of growth. The name Dandelion is a corruption of the French *dents de lion,* referring to the jagged-toothed leaves that can be said to resemble a lion's well-equipped jaw. The bright yellow blossoms are very sensitive to light and weather conditions, closing up before darkness and storms. When the flower head has matured, it closes again with the calyx drawn in a cylindrical shape around the ripening ovaries. It is this phase that gives us the name swine's snout. When the seeds have ripened, the "swine's snout" opens to form the familiar fluffy ball of seeds; each with its own "parachute" waiting to be dispersed by a breeze or a child's breath.

Children all over the world have invented games and predictions that use the Dandelion seed ball. They tell time by the number of puffs necessary to blow off all the seeds. The number of seeds left after one puff reveals everything—the number of children they will have, the number of years until they marry, or the number of sweethearts they will have. The number of questions that the Dandelion seems to answer is as limitless as a youngster's imagination.

After all the seeds have flown, the disc on which they had stood is smooth and bare, with remains of the bracts hanging down beneath. This earned the plant the name priest's crown during the middle ages when a priest's shorn head was a common sight.

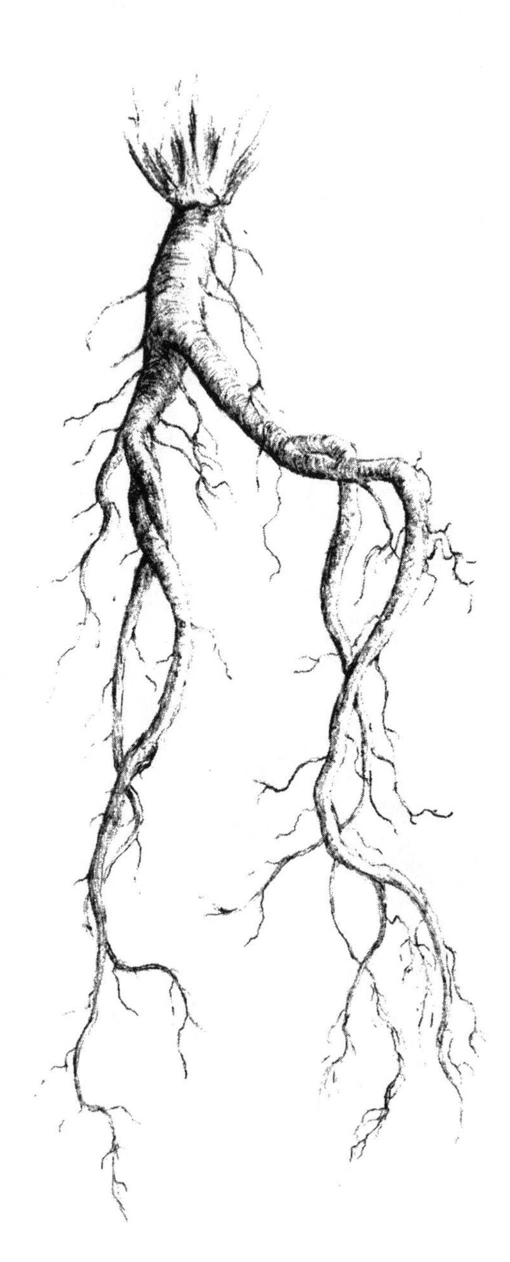

In practical application, it is for spring greens that the Dandelions are most well known. In early spring, before much of the plant world is stirring, the leaves of the Dandelion are at their prime for eating. It seems that after the plant blossoms, its leaves become tough, bitter, and less desirable as a food. However, in the southern and middle latitudes of this country, I have eaten them in late autumn and early winter and found them very agreeable. The best way to prepare a salad of Dandelion greens, or any other sharp wild greens like chicory or wild lettuce, is to use a dressing containing lots of vinegar. The tartness of the vinegar does wonders in cutting the sharp bitterness of the greens. They can also be steamed, boiled, or added to soups.

Many people are very sceptical about the nutritive values of wild plants, even though these same people aren't so concerned about the sprayed and chemically processed foods they normally consume. Although nutritional information on many wild plants is not available, studies have been done on Dandelion greens. They are found to contain iron, phosphorus, sodium, and particularly large amounts of calcium and potassium. They also contain vitamins B and C, and more vitamin A than any garden plant. The only plant that even comes close to rivaling Dandelion's vitamin A content is *Chenopodium album*, the sometimes troublesome but delicious "weed" known as lamb's quarters, or pig-weed.

There is a very efficient technique of collecting Dandelion greens that I didn't learn till well after I acquired a taste for them. Late in the afternoon one spring day I was meandering along the edge of a pasture that I wasn't so sure I was allowed to be in. Picking ingredients for a wild salad, and more or less welcoming the greenery back after a long snowy winter, I looked up to see two big, burly guys drive up and hop out of the proverbial pick-up truck, with proverbial rifle rack in the back window—and walk toward me. The thing that was particularly upsetting was that they were each carrying a butcher knife. I must say, I breathed a small sigh of relief when they both nodded a greeting to me, squatted down and began gathering Dandelion greens!

Their method of gathering Dandelions was to slip the knife under the whole plant and slice it off at the top of the root. This allowed them to gather the whole plant, including the best part, the delicate unopened center called the heart or crown. Of course, some people will argue that the best part is not the crown but the delicious wine made from the blossoms.

The roots of the Dandelion can be roasted and made into a coffee, as described in the chapter on chicory. The roots, when fresh, can be cooked by boiling them in one or two waters, or by adding them to stews.

The infusion or decoction of the dried root has great acclaim as an all-around tonic and blood purifier, being gently diuretic, aperient, and stimulating to the whole system. It is also considered a deobstruent, relieving congestion of the liver, spleen, kidneys and urinary tract. The root was official in the United States Pharmacopea from 1831 to 1926, and remained in the National Formulary till 1965.

Dandelion root can be gathered any time of the year, although in autumn or very early spring the roots are most likely to contain the highest amounts of stored nutrients.

Cichorium intybus

Compositae, Composite family

OTHER COMMON NAMES:
Blue Sailor, Ragged Sailors, Blue Daisy, Blue Dandelion, Coffee-weed, Wild Succory

CHICORY

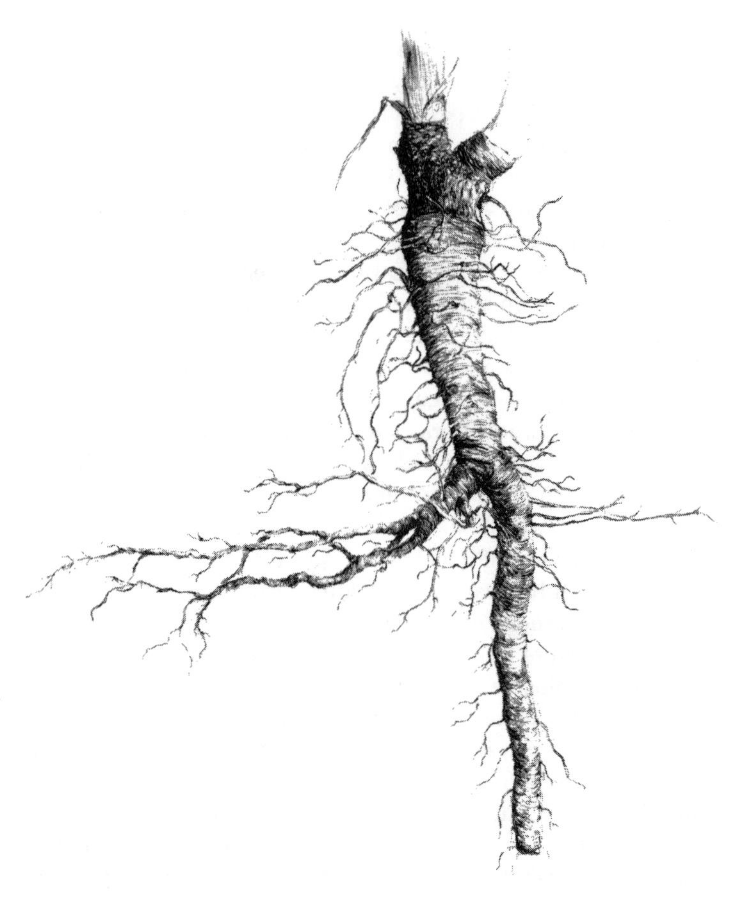

Chicory is one of the many naturalized plants that came over from Europe. It can now be found in fields and roadsides across most of the country.

It is most easily recognized by its one to three foot, angular, erratically branched stalks that bear the bright blue flowers. The flowers are one to one-and-a-half inches across, somewhat resembling a dandelion in shape. The petal rays appear tattered on the outside edge, hence the name ragged sailors. The blooms are usually fully open only in the morning. The leaves are oblong and somewhat toothed, also resembling dandelion leaves in the spring before the flower stalks appear. In fact, they are often gathered indiscriminately with dandelion greens. This presents no problem because the taste and nutritive value of the two plants is almost identical.

For many years in Europe and North America the roasted root of Chicory has been used as a substitute or a neutral adulterant for coffee. Originally, the reason for adding Chicory to coffee was purely economic, since Chicory grows in cooler climates and can be obtained much more cheaply than coffee, which has to be imported from the tropics. In this country, coffee with Chicory added came to be used in the South during the privations caused by the Civil War, when both imported commodities and the money to buy them with were scarce.

Even though the economic need no longer remains, the taste for

Chicory has survived several generations. In many parts of the deep south Chicory is still added to coffee and served as the preferred brew.

I don't drink coffee because I feel that the harshness of the caffiene does me no good. Usually herb teas satisfy me, but now and then (especially on a cold winter's day), I feel the desire for a rich, hearty cup. That's when I reach for the roasted Chicory root. Usually, I have it mixed with roasted dandelion root and a little carob powder to give a hint of mocha. After simmering for a few minutes, it is strained and served with milk and honey. It is one of the most delicious and warming beverages I can name. On special occasions, I steep a little peppermint in it to make sort of a mocha-mint, or serve it with a piece of cinnamon as a flavorful swizzle stick. Although it does have the slightly roasted, bitterish flavor, reminiscent of coffee, it is an entirely different beverage. I call it the *un*-coffee, and most inveterate coffee addicts will agree with me that "it sure ain't coffee." But that's not to say it isn't very pleasurable.

To prepare your own Chicory root coffee, simply scrub the roots well and roast them carefully in a slow oven until they are brown all the way through. This may take a few hours. Then chop or grind them thoroughly and your "un-coffee" is ready to brew.

The unroasted root is used medicinally in much the same manner as the root of dandelion—as a healthful tonic, blood purifier and decongestant of the internal organs. It is a straight, perennial taproot and can be gathered any time of the year.

Ipomoea pandurata

Convolvulaceae, Bindweed family

OTHER COMMON NAMES:
Man-of-the-earth; Wild Potato Vine, Wild Jalap

MAN ROOT MORNING GLORY

Man Root Morning Glory is found in dry, open areas east of the Rockies, from Ontario to Connecticut and south.

The plant is a low growing, trailing vine twelve or more feet long. The leaves vary from heartshaped to fiddle or halberd-shaped. The flowers are funnel shaped, three to four inches in diameter with a deep purple throat.

I had heard lots of stories about some kind of a large-rooted morning glory, and spent a great deal of time and effort looking morning glories in the "eye" to find the purple center. Finally, one hot summer day in western North Carolina, I stumbled upon a purple-eyed specimen. Taking my manual, I sat down beside it for a while to become better acquainted. Three spindly vines ran out of approximately the same area. Soon I decided that I would explore the root to see if it was all that it was supposed to be. As I started to dig, I could see that the stems all originated from a central area about two inches thick (this can be viewed in the drawing at the top of the root), but as I carefully dug deeper around the root, I came to the narrow area around the neck and started to become disappointed because I thought the root was tapering off. Two hours later, as I was lying on my belly midst piles of dirt, the trowel and my hand stretched deep down into the hole, carefully removing dirt from around the root that was still going deeper, I certainly had to chuckle at my pre-judgment. The root was three feet long where I cut it off, and about as big around

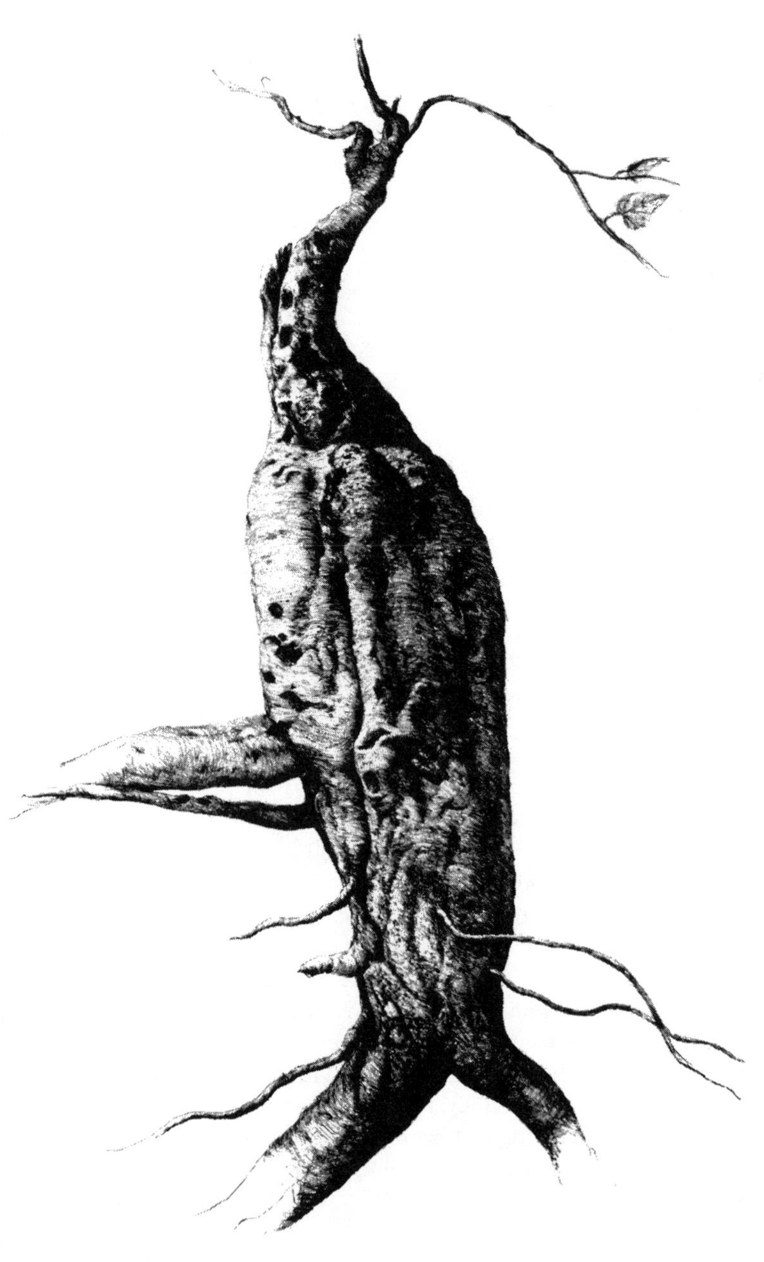

as my thigh. There are reports that some roots have been dug that weighed 40 pounds.

The Indians called the plant *Mecha-meck*, and used the root for both food and medicine. There are modern reports of people who have roasted the root and found it to be somewhat like a "sweet potato but had a slightly bitter taste." I tasted some of the roots in their raw state and they seemed palatable at first, but with an unpleasant, bitter aftertaste. In an attempt to dispell this taste I tried baking it and boiling it for prolonged periods in several changes of water, but the results were always the same. At first bite my hopes were raised, but that insidious "slightly bitter taste" again crept in and prevented me from wanting a second bite.

Man Root Morning Glory was listed in the United States Pharmacopea from 1820 to 1763 as a cathartic (which makes me wonder about its edibility) and as a treatment for various kidney and urinary disorders.

Daucus carota

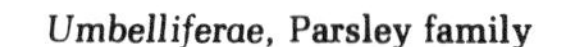

Umbelliferae, Parsley family

OTHER COMMON NAMES:
Queen Anne's Lace, Bird's Nest, Bee's Nest

WILD CARROT

Wild Carrot is found in meadows, fields, and roadsides throughout most of this country, though it was naturalized after importation from Europe. It grows from two to four feet tall. The leaves are compound and finely divided, like carrot tops, and the whole plant is covered with easily visible hairs. These hairs are a very important identification mark because poison hemlock, which is very similar to Wild Carrot in general appearance and habitat, has no hairs. Confusion between Wild Carrot and poison hemlock has been the last mistake of many enthusiastic wild food foragers.

Wild Carrot is a biennial which, from June to September, in its second year produces the intricately delicate flower known to most of us as Queen Anne's Lace. The blossom is a symmetrical umbel consisting of many white florets with one single purple or maroon colored floret in the center. Sometimes this central floret is called "the queen," as she is surrounded by her "lace." Another story tells that the maroon floret commemorates the drop of blood spilt when Queen Anne pricked her finger as she was sewing lace. This head of flowers, when in full bloom, is several inches in diameter and is flat or slightly convex in shape. As the blossom fades and the seeds begin to ripen, the head forms into a hollow cup which gives it the appearance and name of bird's nest. The reason that Queen Anne's Lace is known as Wild Carrot is because it is just that, a carrot gone wild. For the relationship to become clearer, just leave a few of your garden carrots in the ground this fall, and next summer they will form into exact Queen Anne's Lace, identical to the wild ones.

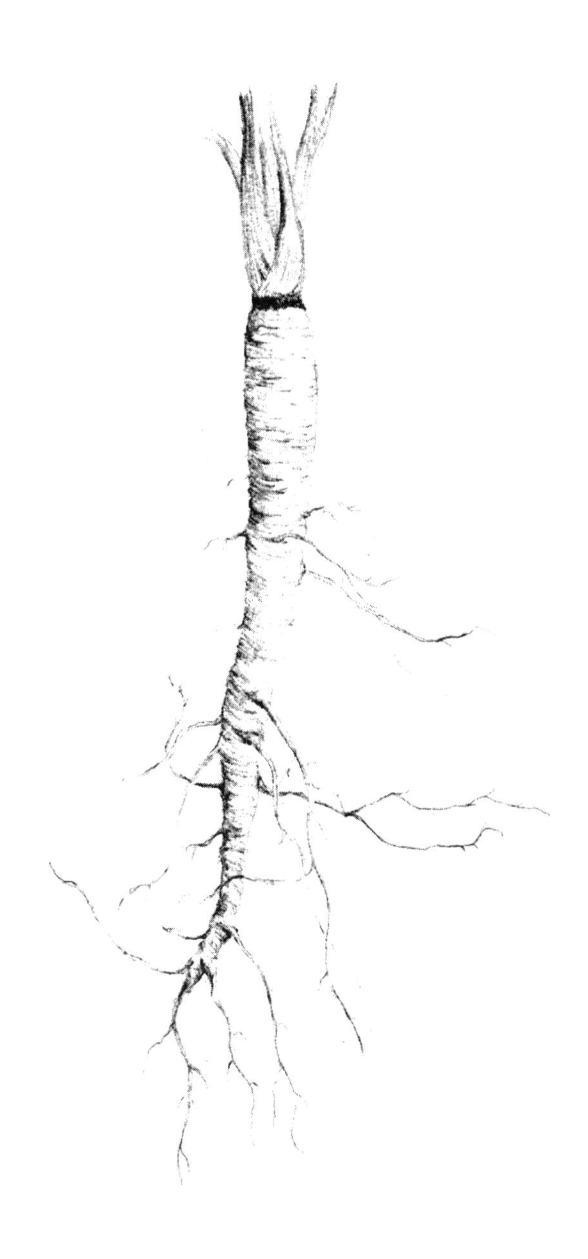

Many generations of living outside of garden gates have forced the Wild Carrot to become lean and tough, and its root has lost most of the color and succulence of its easy living domestic sibling. The smell and taste is still there if you catch plants in the late autumn or winter of the first year, or the early spring of the second year. This, of course, is when the root contains the largest amount of stored food, to be subsequently used for producing the flower and fruit. These whole, thin, white carrots can be boiled or steamed for about fifteen minutes and served with butter. Sometimes the core is a little tough and fibrous, but it will readily slip out when the carrot is cooked, and what remains is good food. I usually ignore the core and slice the root crosswise in thin slices then steam or saute the carrots with other vegetables. Wild Carrots don't have the same tenderness and juiciness as garden carrots but they do have a lot of the same flavor. Cultivating a taste for Wild Carrots is certainly much easier than cultivating a garden full of the domestic ones.

Both the leaves and the seeds can be dried and used as a seasoning, or made into tea. The leaves are best when gathered from the first year plants and the seeds have the most flavor when gathered soon after they are ripe in autumn.

The tea made from these parts of the plant is refreshing and pleasant tasting and that is reason enough for me to drink it. In case you would like some further encouragement, Wild Carrot leaf tea has been used as a general tonic and as a diuretic to treat kidney diseases and bladder problems, and like the seeds, is considered a good carminative to settle the stomach and relieve gas pains. It is comforting to know that if your stomach is upset from overeating the root, the upper portion of Wild Carrot contains the remedy—a complete carrot!

The seeds are also employed as a decongestant for the liver and other organs. The seeds were considered an official drug, and were listed in the United States Pharmacopea from 1820 to 1882.

Saponaria officialis

Caryophyllaceae, Pink family

OTHER COMMON NAMES:
Bouncing Bet, Soaproot, Latherwort, Fuller's Herb, Crow Soap, Wild Sweet William, Bruisewort, Dog Cloves, Lady-by-the-gate, London Pride, Wood-phlox

SOAPWORT

Soapwort is naturalized from Europe and is very common along roadsides and railroad right-of-ways, and in fields and waste places from Maine to Georgia.

Soapwort usually grows one to two feet tall, with opposite leaves. The pink to white, five-petaled flowers are clustered at the top of the plant and somewhat resemble phlox. They are in bloom for almost the entire summer.

Soapwort, as both its Latin and English names indicate, contains large amounts of the chemical saponin, which lathers in water. The name fuller's herb comes from the fact that this plant was used by early New England textile manufacturers to clean and give body to woolen cloth in a process called "fulling". More recently, the plant has been used by museums to restore and clean antique tapestries and other valuable fabrics. The Soapwort plant reputedly surpasses many of the more modern chemical cleaning techniques in bringing out the original depth of the color in these pieces.

I decided that if this plant can be trusted with the task of cleaning precious tapestries with centuries of dirt set into them, perhaps it might also serve as a shampoo for my own curly locks.

Several herbal colleagues and I ran Soapwort shampoo through a rigorous battery of tests under a variety of experimental conditions including everything from skinny dips in a mountain lake to hot showers with a friend. We have concluded that the best way to prepare the Soapwort for shampoo is to gather the whole plant. Put the leaves in a blender with water, and blend them enough to chop them thoroughly. After scrubbing the roots well, break or chop them up a bit. Place them in a pot of water and let them simmer for about fifteen to twenty minutes. (It smells like potatoes cooking.) Add the leaves and their liquid and allow the whole thing to steep. When it is cool it is ready to use, although I usually like to let it set overnight. Be sure to strain it before you use it. It has a slippery, soapy feel, and when poured from one container to another it builds up an impressive head of suds. For many years it was used as an ingredient in beer, for just that purpose. However, it is not nearly as lathery or as concentrated as soap or detergent. To use it as a shampoo, I take at least a quart of it and pour a small portion at a time through my hair, using a brush to help spread it and untangle snarls. I rinse and reapply several times. It feels more like a cream rinse than a shampoo, but it seems to do an adequate job of cleaning, and it really does put a luster in my hair that I don't usually get with regular shampoo. If you are tired of using those expensive "herbal garble" shampoos that have a fragrance about as herbal as cheap perfume, and are as vegetable-colored as a green neon light, this might be an alternative. I have used nothing but Soapwort as a shampoo for months at a time with good results.

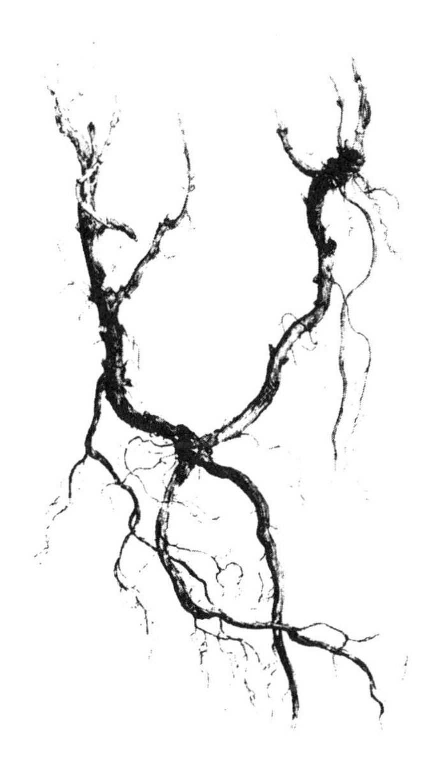

When I prepare the shampoo for myself, I usually boil up a gallon or two at a time so it will be ready to use. It will keep well under refrigeration, or if refrigeration isn't available, vinegar will preserve it. I imagine that it could also be canned in quart canning jars. Just open a jar and you're ready to shampoo.

If you want to add true herbal fragrance to the shampoo, steep aromatic herbs in it when it is hot. Be sure to keep it covered until it cools so the aroma doesn't escape. I usually use rosemary, yarrow, and nettle because these herbs have restorative and tonic effects on the hair as well as a good smell. Since my hair has a tendency to streak blond in the summer, I also have been using a chamomile rinse, which is reputed to lighten fair hair. Since using the chamomile rinse, I am not sure if my hair has actually become any blonder, but I do know that I have been having a lot more fun.

The soothing, cleansing properties of the soapwort solution are used externally for treating skin problems and discharging sores, or as a douche, or a rub for cramps, rheumatism and arthritic joints.

It has also been used for jaundice and other liver problems but because of the saponin content (saponin is toxic), it is not recommended that it be taken internally without qualified supervision.

The roots and rhizomes can be gathered at any time of the year. Once Soapwort gets established, it rapidly spreads by sending out runners from the underground portion of the stems. Even on an early winter's dug root (as in the drawing), the new white runners can be seen emerging at various nodes, in preparation for the warmth of spring. Because of this means of spreading, it is rare to see a solitary Soapwort plant. If you see one, it is usually in a patch with many others.

Hemerocallis fulva

Liliaceae, Lily family

DAY LILY

The Day Lily is another garden plant from Eurasia that jumped the fence and now is widespread in meadows, along roadsides, and around old homesites throughout most of the eastern half of the United States.

This is a familiar midsummer flower that stands on naked stalks two to four feet tall. The flower stalks rise from clumps of long slender curving leaves. Each stalk bears several of the bright orange flowers, each of which blossoms for only one day. On this continent they produce no seed but propagate underground from the root.

The dried flowers are imported from the Orient, and can also be bought in delicacy stores and in Chinese groceries where they command high prices. Or they can be easily gathered out of doors. All stages of the flower are edible. The unopened buds are delicious if steamed for a very few minutes and served with butter and seasoning. For a slightly more festive interpretation, dip the opened blossoms in a rich egg, milk, and flour batter and fry them quickly in hot oil. I remember one such festive flowery Fourth of July dinner when we had elder flower pancakes with rose petal jam and violet syrup, plus tall cool glasses of locust water made by soaking the locust blossoms in cold well water to extract the nectar and floral essence. For greens we had steamed and buttered milkweed flower buds. The main course consisted of fried stuffed Day Lily blossoms. We stuffed them with everything we could think of from rice and cheese to applesauce and even peanut butter! They were all delicious and we stuffed ourselves too. Though in retrospect, the peanut butter was a bit strong, somewhat overpowering the subtle flavor of the Day Lily.

We sat back after this feast sipping chamomile and red clover blossom tea and dandelion flower wine—the silence being broken only by distant fireworks and an occasional floral burp.

The blossom, after its single day of splendor, wilts and remains hanging from the stem for a day or two. These wilted flowers can be gathered and dried easily in a hot, airy attic and stored for use in soups or other dishes. They add a unique pleasant taste and act somewhat as a thickener. The buds and blossoms can also be dried, but because they have more moisture in them to start with, they often mold before they dry in the humid summers we have in the east.

Herb dryers that use artificial heat are very useful, or an oven turned on low, or with just the pilot light on and the door propped open slightly also works well. A car parked in the sun with the windows opened a crack also serves excellently to dry herbs and flowers. Spread the flowers (or herbs) on screens, or lacking screens, newspaper will do.

Let us not forget that the plant owes its existance and hardy persistance throughout its widespread range to its lowly root. The root is fibrous, consisting of a cluster of tuber-like swellings. These are delicious eating. Just boil them for about 10 minutes. They have a crisp texture and unique flavor. The root in the illustration was dug in the late autumn. During the growing season the root clusters are well supplied with a crop of brand new white roots. These are so tender that they need not be cooked. They can be washed, sliced and used as a salad or added to one. They are crisp and have a nut-like flavor. Day Lilies are prolific and it would take a lot of effort to wipe out a bed of them. If you want them to spread, however, just choose a sunny spot where you would like to see them and replant what is left after you pull all the desirable roots off the clump and chances are they will take root and establish themselves in their new surroundings.

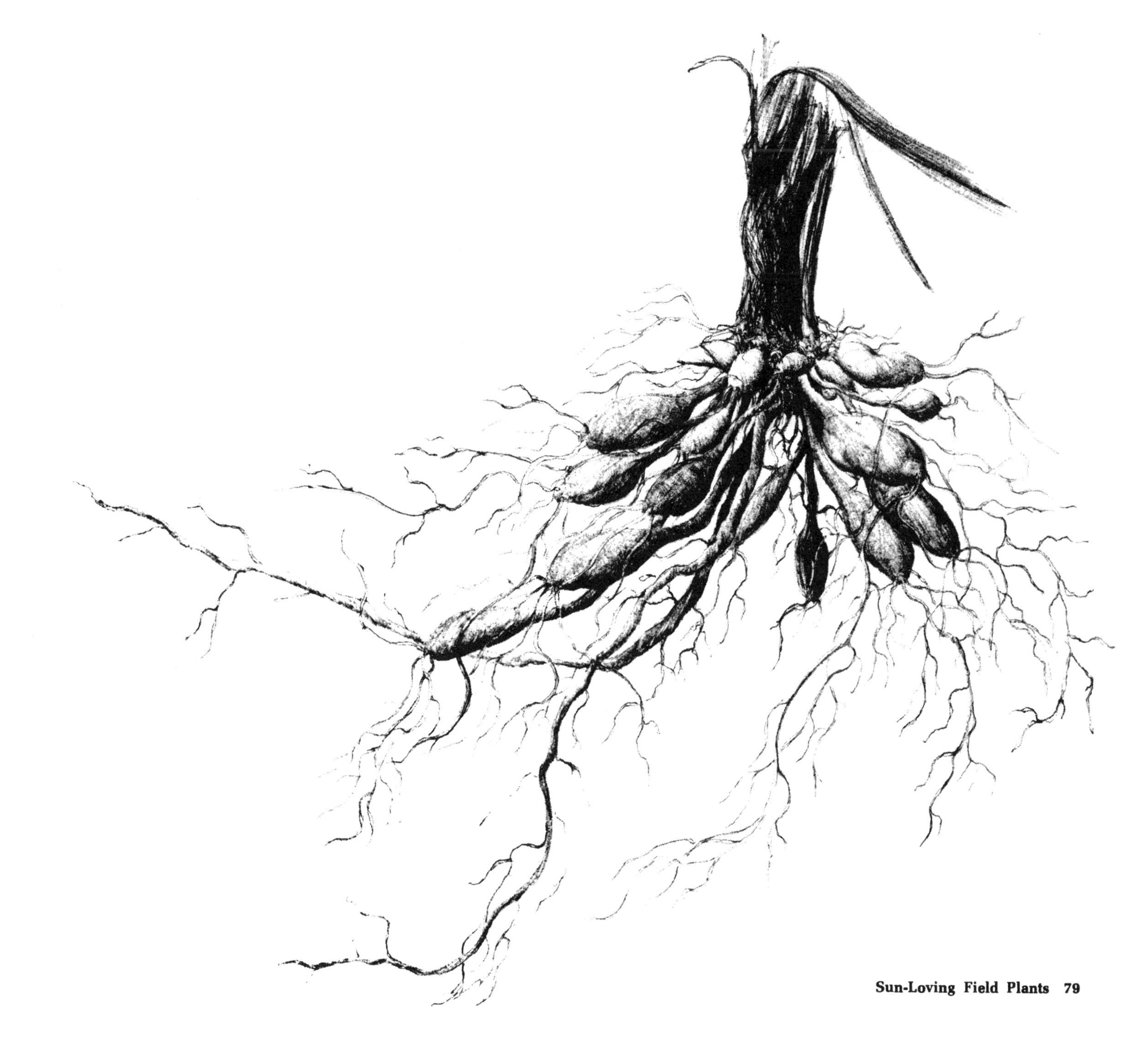

Baptisia tinctoria

Leguminosae, Pea family

OTHER COMMON NAMES:
Horse-fly Weed, Indigo Weed, Rattle Bush, Yellow Broom, Clover Broom

WILD INDIGO

Wild Indigo is found in dry, poor soils in open areas, from Maine to Minnesota south to Florida and Louisiana and grows two to three feet tall with clover-like leaves and brilliant yellow, pea-like blossoms. Many people know it as horsefly weed because of the custom of tying a bunch of it to the harness of a horse to repel flies. One of the most distinctive characteristics of the plant is that the foliage turns dark bluish-black upon drying. Wild Indigo does contain a certain amount of blue pigment but to liberate it is a very complicated process involving fermentation and the use of chemicals which neither the vegetable dyers that I know, or I, have been able to do. The information I have found all states that this pigment is an inferior grade of color anyway. As a medicinal root, however, our native Indigo has had considerable attention and use.

It was official in the United States Pharmacopea from 1831 to 1842, and in the National Formulary from 1916 to 1936, classed as an emetic cathartic, stimulant, astringent and antiseptic. A decoction of Indigo root was used as a remedy for scarlet fever, typhus and epidemic dysentery. It is a stimulant to the liver, intestinal tract and the nervous system and is also used to stimulate the elimination of accumulated wastes in the body, that manifest themselves in various forms of ulcerations and eruptions.

Caution is advised, since Wild Indigo root can be rather drastic in action and large doses have been known to be toxic. Externally, it has been used as an antiseptic for cuts and wounds. In combination with non-alcoholic witch hazel and calendula, it is used to heal cracked nipples of nursing mothers.

Wild Indigo has a shallow woody root that is best gathered in autumn.

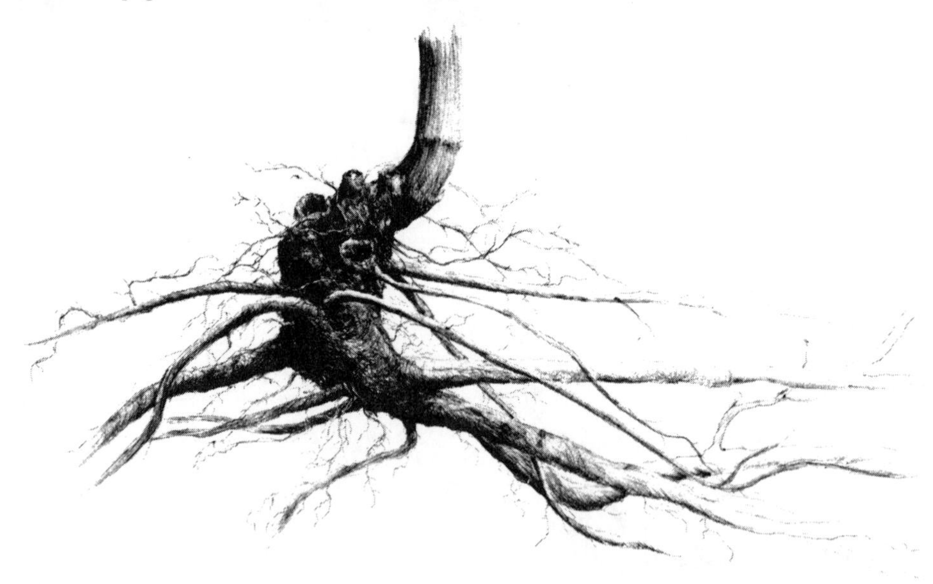

Angelica atropurpurea

Umbelliferae, Parsley family

OTHER COMMON NAMES:
Purple-stemmed Angelica, High Angelica, Masterwort, Archangel, Aunt Jerichos, Alexanders, Bellyache Root

AMERICAN ANGELICA

Angelica is found in most areas of the northeastern quarter of the United States. It is considered a perennial plant because it usually lasts more than two years, but, in fact, is neither annual nor naturally perennial since the whole plant dies off after seeding once. This seeding usually occurs in the third year, rather than the second year as with biennials. I would say that it could be more accurately termed a "tardy" biennial.

It is a handsome plant in its later stages, standing three to ten feet tall with a hollow, purplish, flowering stem. The large compound leaves are divided into threes and sometimes measure two feet across. The greenish white flowers are produced in midsummer in large rounded umbels. The nectar of these flowers has the ability to intoxicate insects that feed on it. It is not uncommon to find butterflies so drunk, from sipping too freely of the Angelica nectar, that they are completely unable to fly. Bumblebees (Bombus spp.) are also well known for their clumsy aerial antics after they have been visiting Angelica flowers. A technically oriented botanist friend of mine refers to such an Angelica intoxicated bee as a "bombed *Bombus*."

The seeds are ripe in late fall. They are spicey and pungent to the taste, and have been used in confectionery and in the preparation of liquors. I find that they add a unique flavor to sauteed vegetables or curry dishes. The stem, gathered while it is still tender, and candied, has long been a popular confection in England and other parts of Europe.

All parts of the plant, especially the root, are useful medicinally and are most valued for the strong aromatic properties that serve to relieve gas pains and generally settle the stomach. Angelica also acts as a general stimulant and tonic and has been used to treat bronchial and other respiratory ailments. In contrast to the flower's effect on insects, it was claimed during the 1800's that medicinal preparation of the root would cause a disgust for spiritous beverages.

The root is a thick taproot similar to those of most biennials. The Angelica root most commonly found in commerce these days is from the European *Angelica archangelica*. There are several species found in this country, most of which have the same properties. The root that I have drawn is from *Angelica triquinata*. The various parts of Angelica have been official in the United States Pharmacopea from 1820 to 1873, and in the National Formulary from 1916 to 1936.

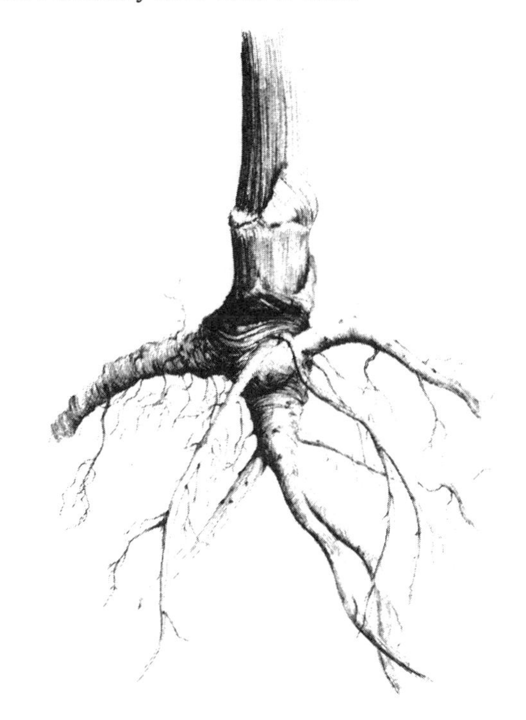

Arctium lappa

Compositae, Composite family

OTHER COMMON NAMES:
Beggar's-buttons, Cockle-button, Clot-burr, Hardock, Hare-burr

BURDOCK

Burdock has been naturalized from Europe and has had no trouble going native in this country. Primarily, just two species are found east of the Rockies; the great Burdock *Arctium lappa* and the common Burdock *Arctium minor*. They are all but identical in appearance and use, the common Burdock being slightly smaller. Great Burdock is found from Canada south to the Carolinas in the cooler and more mountainous parts. The common Burdock, however, is more widespread in the middle and southern states. They both prefer the disturbed ground of barnyards and meadows.

Burdock is a very large, coarse biennial, three to seven feet tall. The leaves are large, heartshaped and wooly underneath; the lower ones sometimes measuring eighteen inches in length. Second year plants produce the tall, branched, somewhat fleshy stem. In the latter part of the summer until frost, the purple, thistle-like flowers are produced in large numbers. The flower heads form the hooked, spiny seed containing burrs that catch and adhere to clothing or the fur of animals. It is these burrs that give Burdock both its widespread range and its reputation of being a pest.

The Burdock has an assortment of edible parts and stages. In early spring the new leaves and stem are edible and make a passable cooked green if boiled in a couple of waters.

The root, however, is my favorite part. Only the roots from the first year plants (the ones without the flower stalk) should be used. As is the character of biennials, the first year is spent storing nutrition (usually in the form of starch) in the thick root. In the second year, the root becomes tough and woody because the plant expends all of the stored energy in producing flowers and fruit.

It is this first year build-up of tender, stored energy that we gustatory botanists want to cash in on. The first year plants are fine any time after they obtain a good size in early summer, but I have had my best tasting Burdock roots in the autumn and early winter. Sometimes they are mild tasting enough to eat raw, but usually they need a little boiling. Some people prefer them boiled in two waters, after which they can be served with butter and seasonings. They make a fine addition to soup, oriental tempura or sauteed vegetables. In fact the Japanese appreciate its flavor so much that they have domesticated it as a cultivated vegetable.

The one problem with this vegetable is that you can't just go out and pull it up out of the ground. It must be dug. The root is usually a foot long and goes straight down, requiring some serious digging. The little hand trowel just won't do. Usually a sharp spade or a post hole digger works well if you just dig down beside the plant and pull the roots into the hole. If you plan your holes well, you can sometimes pull several roots into the same hole. Another trick is to find some Burdock growing at the edge of a bank where it can be more easily dug.

Sometimes it seems as though a lot of work is necessary to obtain this vegetable, but do remember that you didn't have to buy the seeds, or plant it, plow it, fertilize it, or weed around it; all you have to do is harvest it. Besides, what better way could there be to work up an appetite for a delicious meal of Burdock roots!

In the case of Burdock, food and medicine are one and the same. The root was listed in the United States Pharmacopea from 1831 to 1916, and in the National Formulary from 1916 to 1947, where it was considered to be a diuretic and diaphoretic. Burdock, in the tradition of spring tonics, is highly valued as a blood cleanser. The dried root is boiled into a decoction which is taken for all kinds of skin diseases as well as for rheumatism, gout and respiratory problems.

An infusion of the leaves is used to impart strength and tone to the stomach. As a poultice, the leaves are believed to be soothing to bruises, sores and inflamations.

The seeds are used internally in an infusion or decoction to treat kidney problems. When mashed and applied to skin as a poultice, they act as a cleansing agent and their oily nature helps restore smoothness and tone to the skin.

It would seem that within one plant (considered by many as nothing more than a troublesome weed) are the makings for both a grocery and a drugstore.

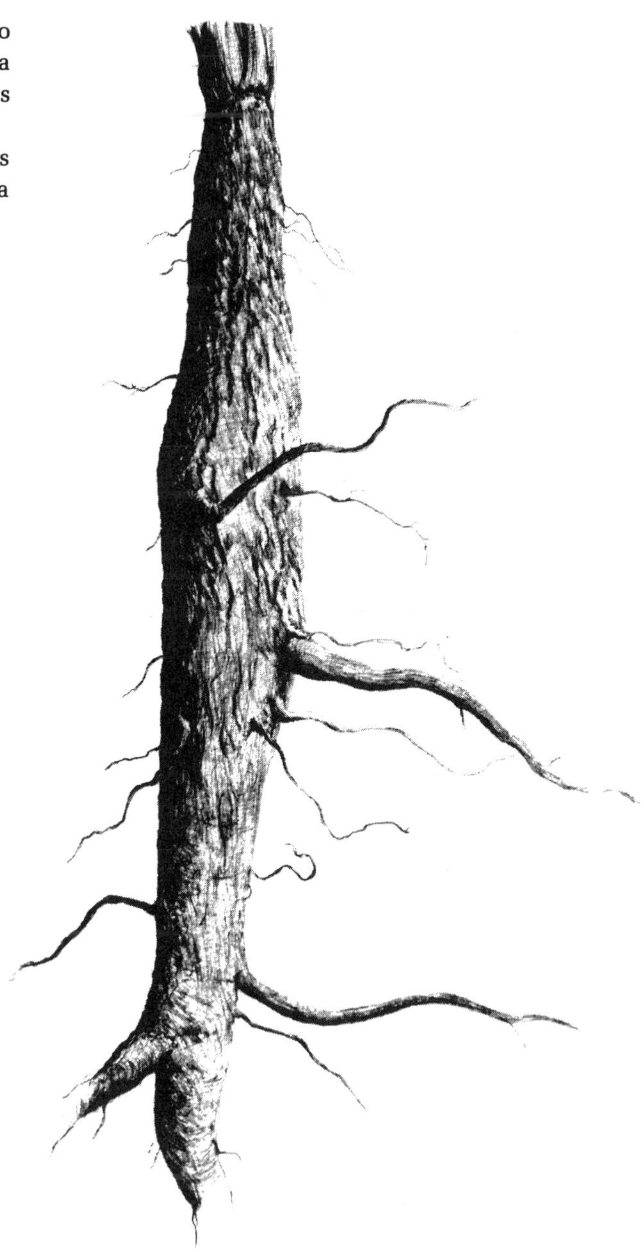

Allium vineale

Liliaceae, Lily family

OTHER COMMON NAME:
Crow Garlic

WILD FIELD GARLIC

Wild Field Garlic grows in fields, meadows and lawns, from southern New England to the Carolinas and into the mid-western states. Its range is shared by several other *Alliums*, and where it leaves off there are other species that take over, so that there is hardly any area in North America where some form of wild onion or garlic is absent. This is a fortunate fact indeed for the foragers and wild food gourmets that abound these days.

The Field Garlic is usually less than two-and-a-half feet tall, and the leaves are somewhat tubular, like chives. The flowers, when present, are pink or white and are arranged in a spherical cluster. More often a cluster of ovoid bulblets are produced instead of blossoms. The bulb resembles a small cultivated garlic and can be differentiated from the wild onion because it is divided into several "cloves" that are flattened on one side. Of course the characteristic that is most distinctive about this whole genus is the odor.

For quite some time I have been trying to clarify the smell and flavor differences between all the wild and cultivated *Alliums*. Even though each seems to have a distinct name, I am never sure where to draw the gastronomic line between garlics, onions, leeks, shallots, spring onions and chives. There must be a gourmet chef somewhere who can enlighten me, but I have yet to run into him (or her).

When I bought my first technical taxonomic plant manual, I thought that this was exactly what I needed. The whole justification for a taxonomist's existance is to establish clear divisions between species. The manuals exhibit incredible amounts of devotion and attention to detail for, even though they may cover only one or two states, they are often thousands of pages long. They describe the differences between almost every green thing.

My expectation was that the literature of a science as sensitive and articulate as this would certainly help me understand the differences between the smells and tastes of onions and garlics. When I flipped through the manual and found the section titled *Allium*, it read, "Perennial, bulbous, scapose, glabrous herbs, usually with *alliaceous odor*." Big help!

Whether you have a wild onion, garlic or leek, there's nothing like that good ole alliaceous odor and flavor to add zest to any wild (or tame) food dish.

Use them raw or cooked as you would any onion or garlic, and use your judgement as to the quantities. The Field Garlic is often considered a nuisance by dairy farmers because sometimes the cows eat it, giving an alliaceous flavor to their butter and milk.

It is the cultivated garlic that is most often referred to in medicine. But the wild garlic probably has the same properties. Garlic has been acclaimed as one of the best internal and external antiseptic healers and cleansers. It is taken as a general tonic and as a standard remedy for everything from the common cold to tuberculosis. It is also used to expell various kinds of parasitic worms. For medicinal purposes, it is best that garlic be eaten raw. To avoid an "alliaceous breath" the garlic can be chopped into pill sized pieces, dipped in honey and swallowed whole.

During World War I, the juice was squeezed from garlic, diluted with water, put on swabs of sphagnum moss, and applied to wounds as a dressing to prevent infection and in this way saved thousands of lives.

All parts of the Field Garlic can be used. The bulblets from the top of the plant in summer are particularly hot and strong and are easily gathered.

Echinacea pallida

Compositae, Composite family

OTHER COMMON NAMES:
Purple Cone Flower, Black Samson-root

ECHINACEA

Echinacea is primarily a plant of the prairies from Michigan to Nebraska south to Alabama and Texas, although it is also found in scattered locations in the east.

The various species of Echinacea, including *Echinacea purpurea* and *Echinacea angustifolia*, are used interchangeably. Since the pale purple cone flower is the one I first came across and illustrated, it is the one I describe here. The plant stands about two feet tall with long, slender, roughly hairy leaves. The flowers bloom from mid-summer into early autumn looking similar to large, droopy, pale purple, petaled daisies. As the flower matures, the center becomes cone shaped. The root of Echinacea was in great favor with the plains Indians as a remedy for many ailments including snakebites, insect stings, headaches, stomach cramps, toothaches, enlarged glands as in mumps, sore throat, hydrophobia and as a treatment for distemper in horses. The Indians also discovered that the plant seemed to enable the body to endure extreme heat, and it was taken prior to sweat baths, and such ritual feats as immersing hands in scalding water or holding live coals in the mouth.

In the white man's medicine, the roots of Echinacea were official in the National Formulary from 1916 to 1950, and have been recognized as an excellent internal cleansing agent and detoxifier. Echinacea has been used to treat blood poisoning, typhoid fever, gangrene, syphilitic conditions, and all kinds of abcesses such as boils, tumors, infected sores, etc.

The root is the deep taproot typical of many prairie plants that must survive both heat and arid soils.

When the Echinacea root is dry, the wood has a peculiar grey, streaked appearance, very little smell, and a strange sweetish taste that turns acrid and is slightly numbing. Chewing a piece of the root stimulates the saliva flow. It is usually collected in autumn.

Valeriana officinalis

Valerianaceae, Valerian family

OTHER COMMON NAMES:
Cat's Valerian, St. Georges-herb, Setwell, All-heal, Capon's Tail, Vandal-root

VALERIAN

Valerian was brought to this country from Europe to be cultivated in the early settler's herb garden. It has no trouble escaping, and now graces meadows and dooryards from Maine to the midwest and south to New Jersey.

Valerian is from two to six feet tall. The opposite leaves are sparsely placed on the stem. Each leaf is divided into many deeply dissected segments. The flowers blossom between June and August and are white with a slight rose blush. They appear as clusters at the top of the plant and are very fragrant.

Valerian root has had a long and active medicinal history. To ancient physicians the plant was known as *phu* or *fu*, which sounds like an apt reaction to the odiferous root. By the tenth century, it was called Valeriana. In some early medical works of that period descriptions of it began *Fu id est valeriana*, until eventually the first name was omitted.

Valerian is a powerful nerve tonic, carminative and antispasmodic. It is used as a nerve sedative that allays pain and promotes sleep without narcotic after-effects. During World War I, in Europe, Valerian was commonly prescribed to civilians under bombardment and suffering from severe nervous strain. Large doses can cause headache, mental agitation, delusions and extreme restlessness. Hilter, along with his other problems, was also reported to be a Valerian addict.

The peculiar scent of Valerian root is particularly attractive to rats and has been used to bait traps for them. Some have suggested that Valerian root was the secret of the Pied Piper of Hamelin's irresistable power over rats. Cats also often respond to Valerian as they do to catnip.

In proportion to the size of the rest of the plant, the perennial rhizome is suprisingly small. When it is freshly removed from the ground it has almost no smell which, upon drying, becomes strong and unmistakable. My experience in harvesting and digging Valerian roots in eastern Maine one summer taught me a valuable lesson in the economics of commercial root foraging. I found a meadow full of Valerian near the coast and a friend and I decided that we would spend a day seriously digging it. The meadow had a thick turf which had to be cut through with a shovel. I found that I could roll back the layer of sod, pull it apart and extract the Valerian which were tangled in among the other roots. It was hard work and we broke one shovel, but by the time the day was over we had a shopping bag nearly full of roots. When we returned home, it took another half day to wash and sort them. Several days later, after the roots were completely dry, I weighed them to find that what we had gathered came to a little more than three pounds, worth about six dollars on the wholesale market. That means my friend and I earned three dollars each—not bad for a day-and-a-half's work! I guess I'm in this business for love rather than money.

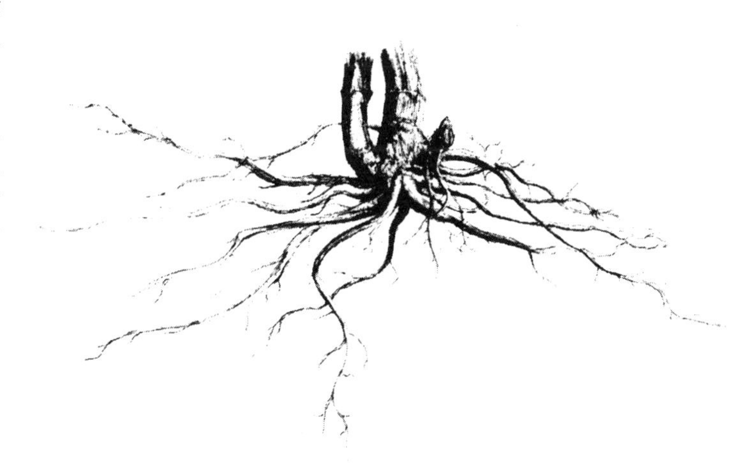

Conium maculatum

Umbelliferae, Parsley family

OTHER COMMON NAMES
St. Bennet's-herb, Poison Parsley, Spotted Parsley, Snakeweed, Wode-whistle

POISON HEMLOCK

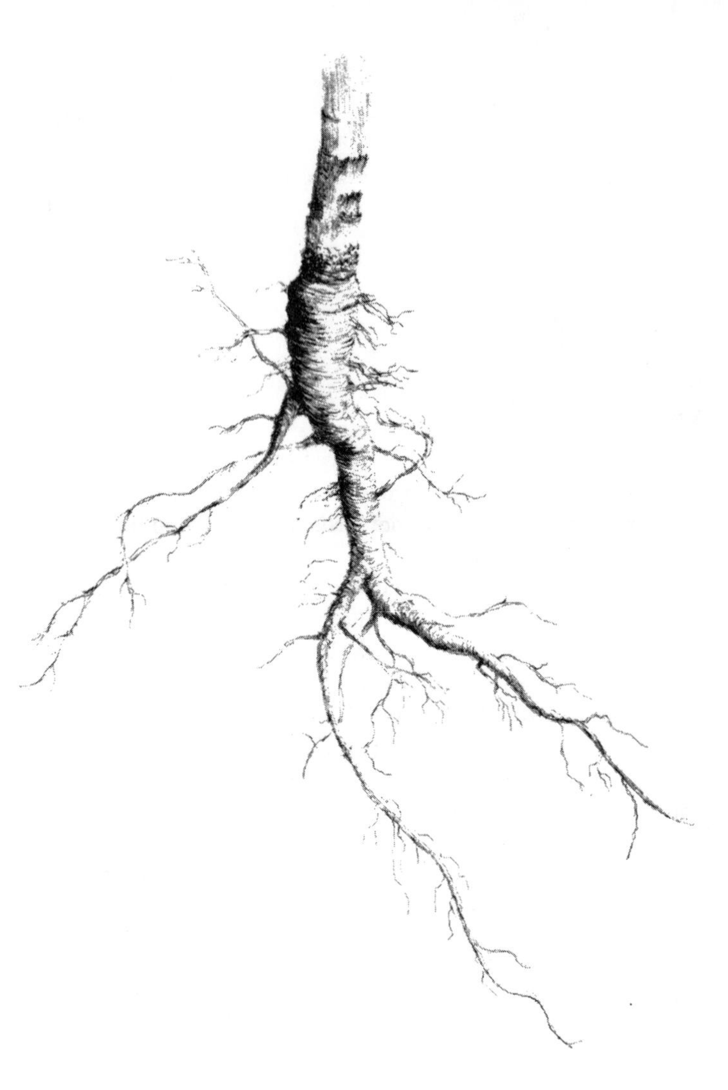

Poison Hemlock has been naturalized from European origins and can be found in moist waste places from southern Canada into Mexico.

The plant is a biennial, often growing to six feet tall. The finely divided, compound leaves resemble those of the carrot except that they are hairless. The stalk is also hairless. It is hollow and often blotched with purple when older. These purple markings, according to an old English legend, represent the brand put on Cain's forehead after he had committed murder.

The dried hollow stalks have been used by children to make whistles and pea shooters. Even though most of the poison has dissipated by the time the stalk has weathered and dried, this practice can still have serious results and should be strongly discouraged.

The delicate white flowers are in small umbel-clusters which give way to smooth seeds that resemble anise or fennel seeds in arrangement and appearance. The elegant nature of this plant does very little to betray the deadly power contained within.

The poisonous property occurs in all parts of the plant, though it is less strong in the root. I have included the root here because the danger of confusing it with wild carrot or parsnip cannot be over-emphasized. There are several close relatives of Poison Hemlock which are nearly identical to it and also poisonous. The root I have illustrated is of small hemlock or fool's parsley (*Aethusa cynapium*). Poison Hemlock is in no way related to the hemlock tree *(Tsuga canadensis)* which is not known to be toxic.

Because of Poison Hemlock's peculiar depressant or sedative action on the functions of the central nervous system, it has been used medicinally to treat cases of undue motor excitability as in spasmodic coughs, certain forms of epilepsy, asthma, early states of Parkinson's disease, and delirium tremens.

It is interesting to note the relationship between Hemlock and strychnine, two potent plant poisons. Hemlock poisons by depressing and paralyzing the motor nerve centers. Strychnine is a motor excitant and poisons by over-stimulation of the nerves, muscles, respiration and heart. Though they are both lethal poisons their properties are antagonistic to one another and each is used as an antidote for the poisoning caused by the other.

The process of treating Hemlock poisoning involves using emetics, and lavage to clear the stomach of any undigested Hemlock; then administering tannin and stimulants such as coffee, strychnine or atropine. The patient must be kept warm and artificial respiration given if necessary.

The most famous use of Poison Hemlock was by the ancient Greeks as a humane method of capital punishment. It is quite painless and the recipient's mind remains clear to the end—a very appropriate way for a philosopher to leave this world. We are indebted to Plato for his early account of the effects of Hemlock on Socrates.

> Then Crito made a sign to his slave, who was standing by, and the slave went out, and after some delay returned with the man who was to give the poison, carrying it prepared in a cup. When Socrates saw him, he asked, "You understand these things, my good sire, what have I to do?" "You have only to drink this," he replied and to walk about until your legs feel heavy, and then lie down, and it will act of itself." With that he handed the cup to Socrates who took it quite cheerfully. Socrates, without trembling and without any change of feature, looked up at the man with that fixed glance of his and asked, "What say you to making a libation from this draught. May I, or not?" "We only prepare so much as we think sufficient Socrates," he answered. "I understand," said Socrates. "But I suppose that I may, and must, pray to the Gods that my journey hence may be prosperous: that is my prayer, be it so." With these words, he put the cup to his lips and drank the poison quite lively and cheerfully. Till then most of us had been able to control our grief fairly well; but when we saw him drinking and then the poison finished, we could do so no longer: my tears came first in spite of myself, and I covered my face and wept for myself; it was not for him, but at my own misfortune in losing such a friend . . . "What are you doing, my friends?" Socrates exclaimed . . . "I have heard that a man should die in silence. So calm yourselves and bear up."
>
> When we heard that, we were ashamed, and we ceased from weeping. But we walked about until he said his legs were getting heavy, and then he lay down on his back, as he was told. And the man who gave the poison began to examine his feet and legs, from time to time. Then he pressed his foot hard and asked if there was any feeling in it; and Socrates said, "No:" and then his legs, and so higher and higher, and showed he was cold and stiff. And Socrates felt himself, and said that when it came to his heart, he should be gone. He was already growing cold about the groin, when he uncovered his face which had been covered, and spoke for the last time. "Crito," he said, "I owe a cock to Asclepius: Do not forget to pay it." "It shall be done," replied Crito. "Is there anything else that you wish?" He made no answer to this question; but after a short interval, there was movement, and the man uncovered him, and his eyes were fixed. Then Crito closed his mouth and his eyes. Such was the end, Eshrecrates, of our friend, a man, I think, who was the wisest and justest, and the best man that I have ever known. (8)

Sassafras albidum

Lauraceae, Laurel family

OTHER COMMON NAMES:
Cinnamon Wood, Smelling Stick,
Saxifras, Ague Tree, Saloop

SASSAFRAS

Sassafras is found from southern New England to Michigan and south to Florida and Texas. It appears in a variety of environments but it prefers clearings and the edges of fields, where it gets a good dose of sunlight. In the northern part of its range, Sassafras is a small spindly tree, but in the South it can be a sizeable tree more than 80 feet tall with a trunk two or more feet in diameter.

The leaves have three distinctive variations in shape, all on the same tree. One is a simple oval leaf. The other is mitten-shaped. The third has three finger-like lobes. Often these three shapes can be found on the same branch. The green colored bark of the young branches and the upper portions of the trees make it easy to identify Sassafras. Clusters of tiny, yellowish-green flowers can be seen in early spring, just as the leaves are beginning to form. Sometimes, I toss a few of these flowers or young Sassafras leaves into a spring salad. They impart a unique flavor and slight mucilaginous quality. In the fall, the female flowers produce dark blue, pea-sized berries that dangle from red stems.

Sassafras is little known for the role it played in the settling of the New World. It was first introduced in Europe in 1560 by the Spaniards who obtained it from the Indians in Florida. It soon acquired a great reputation for its healing virtues. As this reputation spread from Spain to northern Europe, English ships were dispatched to the new colony of Virginia to collect Sassafras, which became the first cash "crop." For many years it rivaled even tobacco as the major export from North America. For example, in 1622, the Colonists received an order from the Earl of Southhampton "for the sending home of three score thousand weight of Sassfras." In an attempt to comply with the order every man was given a quota to bring in, with a fine of tobacco if he didn't meet his quota.

The "exotic" Sassafras tea was becoming very popular in Europe and was served in the most fashionable circles. The trade in Sassafras was so brisk that by 1625 the complaint was made that the London investors "bestow their moneyes . . . upon two commodities onely, Tobacco and Sassafras matters of present profit, but no wayes foundations of a future state." (9)

It was during this period of popularity that Sassafras was not only considered a fashionable, health-giving tonic, but it also came to be known as a veritable cure-all for practically every disease known to man. (It had been listed in the Pharmacopea Londinensis since 1618.) Then it acquired the reputation of a blood cleanser specifically used to treat syphilis. Whereupon its fashionability (if not its use) declined. This factor, coupled with the flooding of the market, caused a drop in the price of Sassafras. Shortly thereafter, the new fad of Tobacco consumption took its much more addictive hold on Europe and relegated Sassafras to a secondary role in the economy.

Sassafras tea is still surviving as a favorite among many people and is certainly a favorite of mine. To make it, all that needs be done is to dig or pull up some of the roots and wash them thoroughly, chop or whittle them small enough to fit into a pot, and boil them.

The water turns from light orange to deep red as the tea is brewed. The strength desired is a matter of personal taste. I like it about medium orange. Sassafras is one of the few aromatic herbs that is boiled to best release its flavor. I find the tea very warming and stimulating when hot. As an iced tea, it is very cooling and refreshing. For variety, I like to steep a little mint in it before it cools.

Sometimes older people object to Sassafras tea saying that it tastes like medicine or tooth powder. Actually they have it backwards. Sassafras oil was commonly used as a flavoring for various nasty tasting children's patent medicines in the early part of the

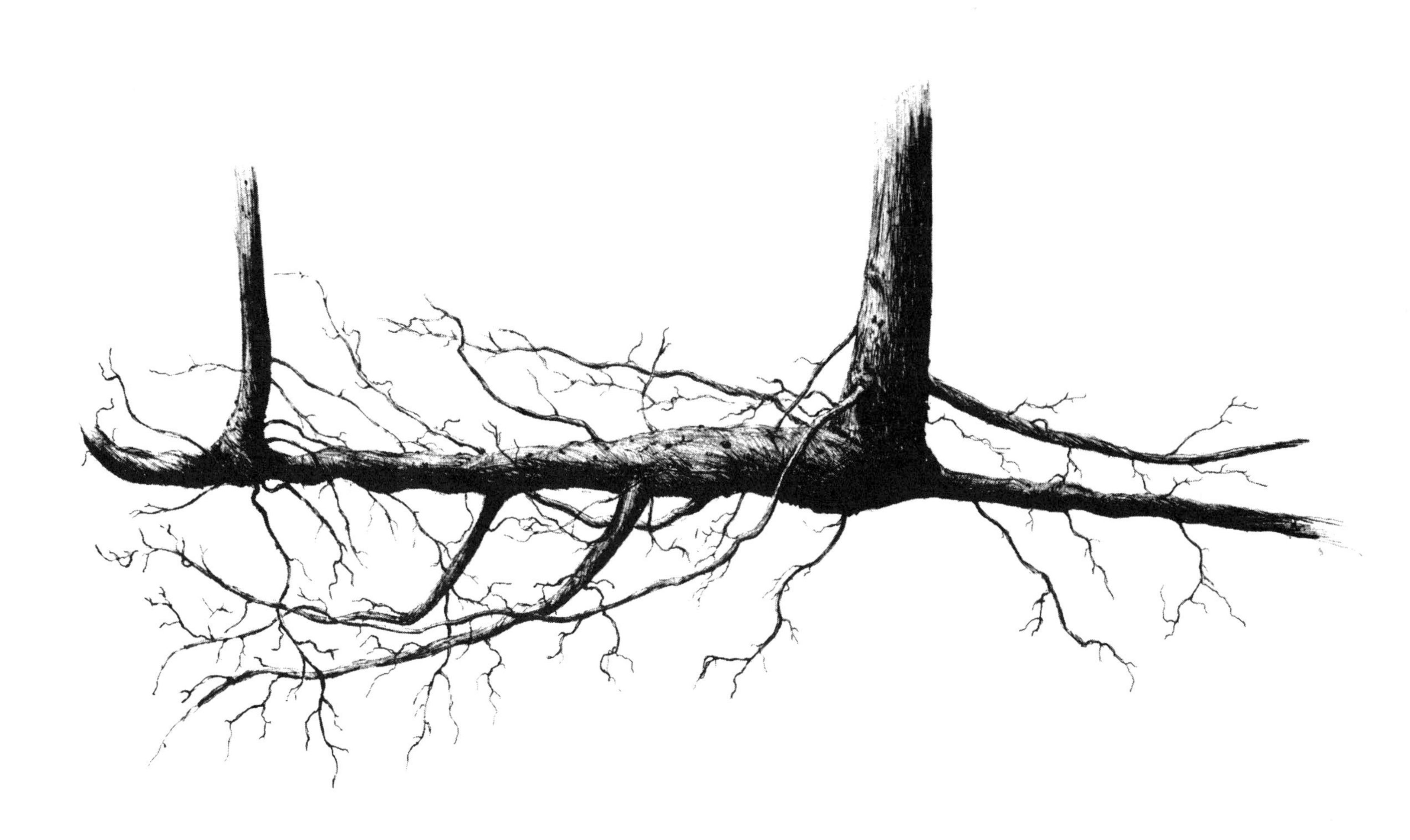

20th century; so it is actually the medicine that tasted like Sassafras. Those of us who were born after World War II were spared the majority of these remedies. This is fortunate, since the active ingredient in many of them was a narcotic opium derivative. Anyway, most young people these days seem to prefer their opium unflavored.

It was from the Choctaw Indians that the French settlers in Louisiana learned to use powdered Sassafras leaves to thicken and add flavor to gumbo soup. These powdered leaves known as "gumbo filé," are one of the basic ingredients in Creole cookery so admired today. If you don't have a bayou in your backyard, or you live in the East, there's still a good chance you have Sassafras nearby. All you have to do to make your own gumbo filé is dry young Sassafras leaves till they're crisp, and powder them in a blender. Then sift them to remove all the stems and large pieces. Place the resulting powder in a shaker and add it to any soup or gumbo just as it is being served.

The bark of the root was official in the United States Pharmacopea from 1820 to 1926, and in the National Formulary from 1926 to 1965. The Sassafras pith, the oil of Sassafras, and safrole, a derivative of the oil, were also listed at various times. The tea and various extracts of the root have been used to purify the blood in skin diseases, rheumatism, and syphilitic problems. The tea, both taken internally and applied externally has been used as a remedy for poison ivy. It has also been a favorite spring tonic to "thin the blood."

Recently there has been a scare because it was found that safrole, an extract of Sassafras oil, can be carcinogenic and cause liver problems. The tea itself has not been found to cause any problems in its centuries of use. I plan on drinking the tea (moderately, of course) at least until the age of 102, at which point I might reconsider the issue.

The pith of the Sassafras is mucilaginous and is used as a soothing demulcent application for irritations and eye afflictions. Sassafras root is also used as a dye.

The traditionally accepted time of the year for gathering Sassafras roots is during the months when leaves are not on the tree and the sap is in the roots. I have found, however, that the season seems to make little difference. Whenever the root is dug, it always seems to have the same good flavor and aroma. I usually gather Sassafras roots from small trees, less than two inches in diameter which can be found growing in thickets. Frequently, the root goes down several inches into the ground and makes a sharp right angle. This makes it difficult to pull, but a well placed shovel can help dig along the horizontal root and make for more efficient gathering. Sometimes one root produces several trees (as in the illustration). If I unearth a big tree or find one already uprooted, I usually just peel the bark from the root, since it is the bark that contains almost all the flavor, aroma and medicinal properties.

Sassafras is one plant you don't have to worry about exterminating. It is considered a weed tree by many, because it spreads very rapidly and almost any portion of the root that is left in the ground is capable of sprouting a new tree.

Asclepias tuberosa

Asclepiadaceae, Milkweed family

OTHER COMMON NAMES:
Butterfly Weed, Orange
Swallow-wort, Orange Milkweed,
Wind Root

PLEURISY ROOT

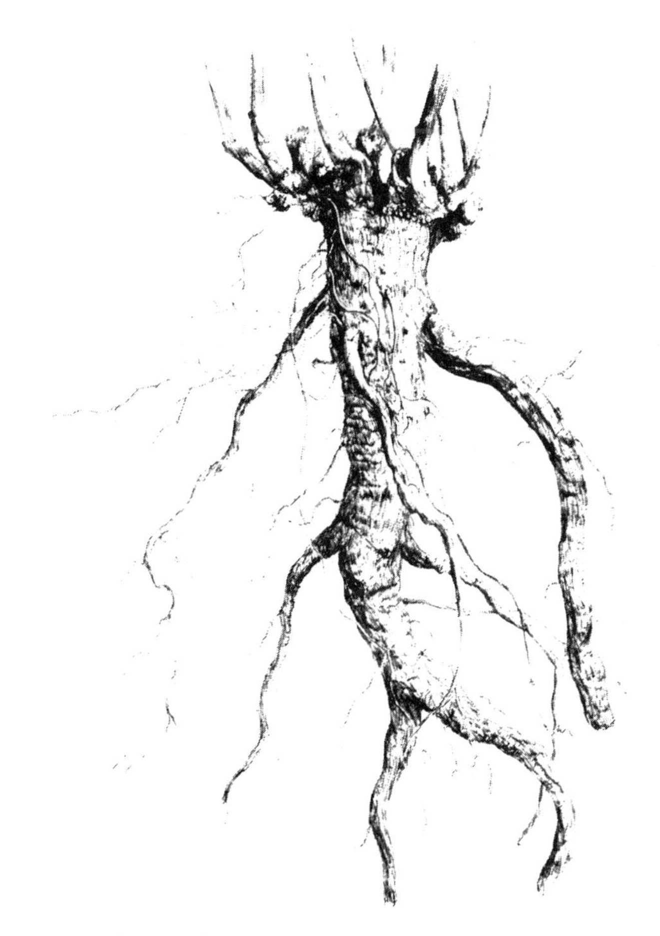

Pleurisy Root grows in the poor soil of dry fields and woods' edges, from southern Canada to Florida and west as far as Arizona. It is most abundant in the central and southern portion of its range.

Pleurisy Root is the most brilliant and showy member of its family. The stiff, hairy stems are about one or two feet tall, bearing a thick growth of coarse, oblong leaves. The bright orange flower heads are borne on the top of the stems. The flowers have the recurved petals typical of the milkweed family, although this is one *Asclepias* that doesn't have much of the characteristic milky sap. The flowers are attractive to butterflies as well as to humans. The plant is in bloom for most of the summer season, after which four to five inch pods appear containing seeds with long silky hairs.

The above-ground parts of Pleurisy Root, unlike the common milkweed, *Asclepias syriaca*, is not edible. The milkweeds can be a tricky genus. The shoots, flower buds, and young pods of several of the most common species are not only edible but are delicious. Some species, however, especially in the West, are poisonous. The root itself can also be toxic in large doses, but in mild quantities can be a very beneficial medicine.

Like its name indicates, Pleurisy Root has been a longtime favorite remedy in a decoction for pleurisy, pneumonia and other pulmonary diseases. It is also used for disturbances of the gastro-intestinal tract and for rheumatism. It acts as a stimulant and expectorant to break up mucous build-up and clear breathing passages. I know of some cases where Pleurisy Root has significantly aided asthmatics. Its diaphoretic powers make it useful in breaking fevers by promoting perspiration.

The surprisingly thick taproot penetrates deep into the ground to better absorb moisture and nutrients. It also acts as a storage reservoir which enables the plant to survive in harsh dry soils. There is another benefit to having this kind of root structure. The plant is usually found so deeply rooted in such hard soil that it is fairly well protected from all but the most enthusiastic herbalist or root-digger.

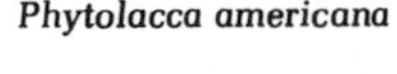

Phytolacca americana

Pytolaccaceae, Pokeweed family

OTHER COMMON NAMES:
Poke, Pigeon Berry, Red Inkberry, Scoke, Pocan, Coakum, American Nightshade

POKEWEED

Pokeweed can be found in moist, disturbed soils along field borders, fence rows and wood edges, from New England to Minnesota and southward to Florida and Texas.

Poke is a large plant from three to nine feet tall, with large, soft, oval leaves. The stems are thick and fleshy and acquire a reddish color as they become older. The bland, greenish-white flowers are arranged in drooping spikes which, in late summer, give way to drooping clusters of distinctive, rich, purple berries containing a beautiful, magenta-colored juice.

The shoots and tender plant tops are a favorite springtime pot-herb of country people. Poke is native to America, but the people of southern Europe imported it long ago. It is now cultivated not only as a garden vegetable, but also an "exotic" ornamental with its showy red stems and attractive fruit. In Europe it is known as Phytolaque, Raisin d' Amerique, Morelle a Grappes, Herbe de la Laque, Amerikanische Scharlachbeere, and Kermesbeere.

The young shoots are commercialy canned in some parts of the American south where they are known as "poke salet." Many novices (and Yankees) hear the name as poke "salad," and are very disappointed by the taste of raw poke in a salad. The term "salet"usually means a wild green to be cooked, most often in the southern tradition of overboiling with salt, pepper and hog fat.

Pokeweed lends itself well to this cooking style. Its strong taste is best tamed by boiling for a few minutes, casting off the first water, and then simmering slowly for about 15 minutes, adding a little oil, butter (or hog fat) and seasonings to taste.

The root is the source of a powerful drug called phytolaccin, which can be toxic in large doses. As spring moves into summer and the plant matures and starts to flower, the phytolaccin moves from the root to the rest of the plant and this makes the plant unsafe to eat during the remainder of the season. This could be Nature's way of protecting the plant during its reproductive cycle.

As an external medicine, Poke root is used in a decoction as a wash or made into an ointment for various skin diseases such as eczema, ulcers, scabies, ringworm and other fungus infections. The root of *Phytolacca* was official in the United States Pharmacopea from 1820 to 1916, and in the National Formulary from 1916 to 1947, where it was classed as a slow emetic, purgative and alterative.

It has been used, in small doses, as an alterative to stimulate the metabolism and to help break up congestion in the alimentary canal, as well as in various organs including the lymph glands. It has also been used to treat breast cancer, and the excessive swelling of breasts after childbirth which sometimes makes nursing impossible. It has often been a part of the formulas used in treating arthritis and rheumatism.

The Poke plant has a large fleshy taproot which can be as thick as a man's arm; but because it usually grows in fine, sandy soil, it is not difficult to dig. After the first freezing weather in late autumn, the top six inches of the root can be chopped off and planted in a box of earth in a dark cellar. The crowns of the roots bear dozens of tiny buds. With frequent watering, these buds will develope into shoots. After these first shoots are cut, they are regularly replaced by a succession of equally strong new shoots which can make a welcome fresh, spring vegetable addition to a mid-winter diet.

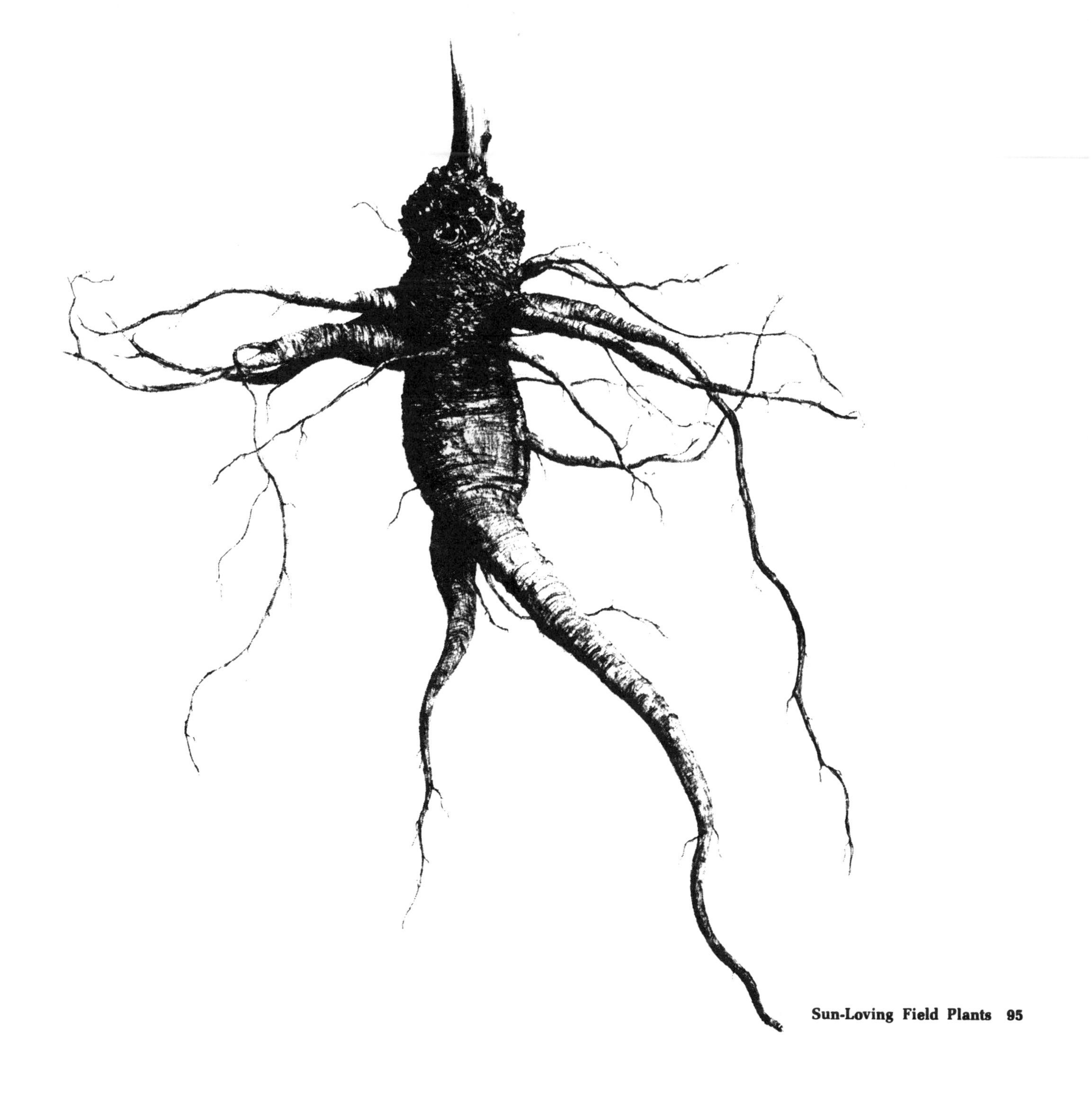

Helianthus tuberosus

Compositae, Composite family

OTHER COMMON NAMES
Sunflower Artichoke, Ground Artichoke

JERUSALEM ARTICHOKE

Jerusalem Artichoke is a sunflower, not native to Jerusalem nor closely related to the artichoke family. Other than that, I guess we could say it is pretty well named. Originally it was indigenous to the central portion of North America, but due to extensive cultivation by both Indians and white men, it now has naturalized itself to most temperate areas of North America, and parts of Europe and Asia.

Its name, Jerusalem Artichoke, is a corruption of the Italian *Girasola articiocco;* Girasola referring to the sunflower's reputation of always turning to face the sun.

During late summer or early autumn, this perennial sunflower can be found in full blossom at the edges of fields, along roadsides and railroad right-of-ways, wherever there is fairly rich, light soil. The plant usually stands from six to more than ten feet tall on single slender stalks with coarse, spearhead shaped leaves up to ten inches long and three to five inches across. The stem is covered with stiff, almost prickly hairs. The coarseness of the leaves and the prickly hairs on the stem, for me, are the best identification points short of taking a shovel to the roots. The flower, like the flowers of most wild sunflowers is about two or three inches across and looks more like a yellow daisy than the large, seed-packed head of the cultivated seed-sunflowers.

The tubers attain their largest size, and should be gathered in late fall after the plants have died back. They are about the same size as a small to medium potato, and have the same food value as a potato, but instead of containing starch they contain an allied substance called inulin, which makes this tuber a valuable food for diabetics and others who need a low starch diet.

Jerusalem Artichokes can be eaten raw, pickled, or cooked any way you would a potato. They have a distinctive, somewhat sooty flavor, slightly reminiscent of a potato with a more watery texture.

In the last part of the seventeenth century, Parkinson, an early herbalist-writer, called them "Potatoes of Canada," because the French first brought them to Europe from Canada. He described them as a "dainty fit for a queen."

When first introduced into Europe, the standard method of preparation was to boil them until tender, after which they were peeled, sliced and stewed with butter, wine and spices—not a bad start for standard cookery. After the initial introduction, the Europeans became more adventurous and baked them into various mince-type pies, with "marrow, dates, ginger, raisins and sack." To switch back to this century, and this side of the Atlantic, Euell Gibbons published a recipe for the ultimate artichoke adventure, Artichoke Chiffon Pie. (10)

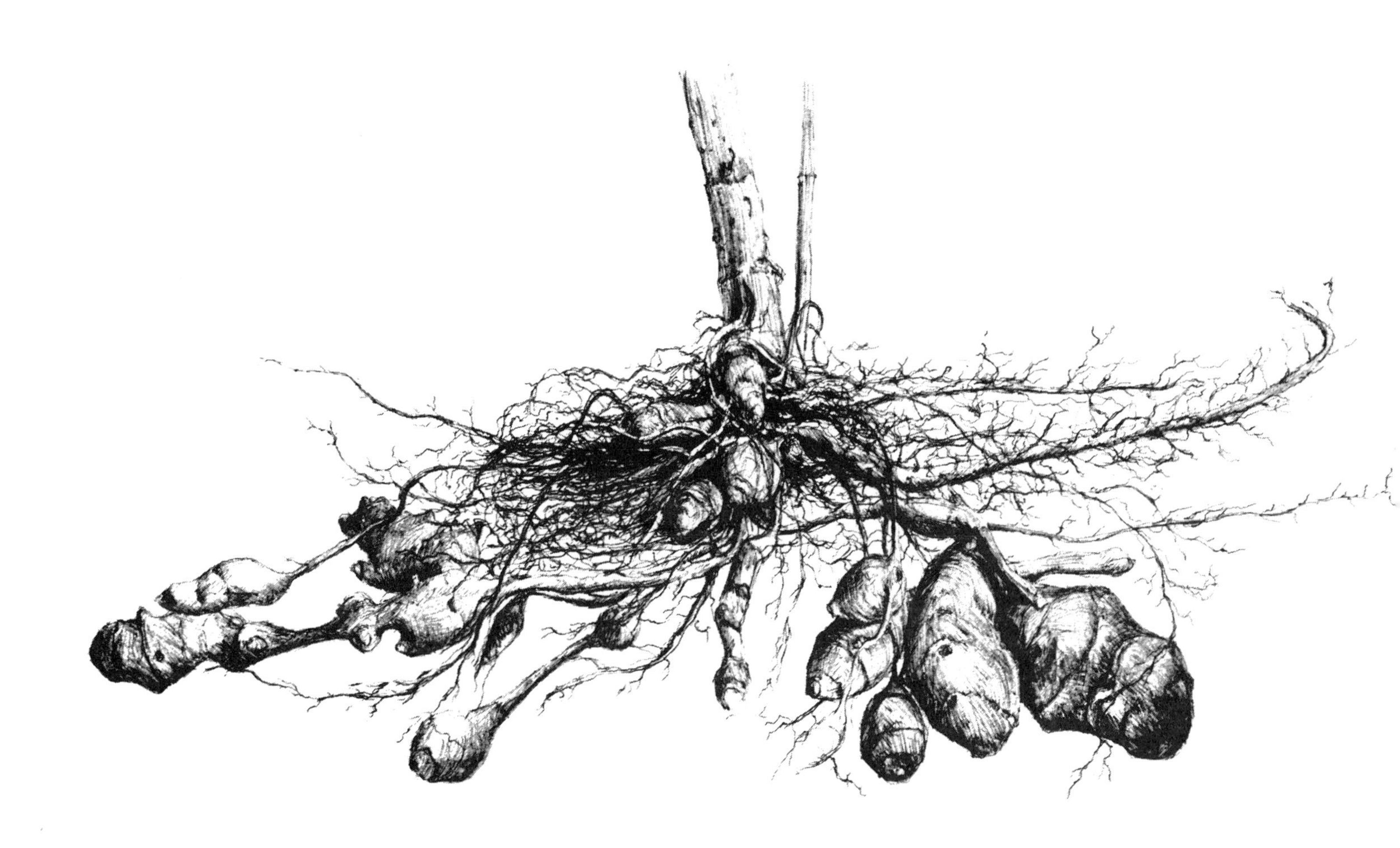

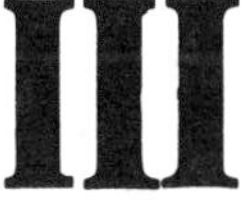

Roots of Aquatic and Marsh Plants

The marsh plants are sometimes characterized by creeping rhizomes. Those that actually grow in water have special porous passageways that supply the roots with air. Because the roots I have selected for this section grow in such a diverse variety of wet conditions, there are very few generalizations that can be made about them. They shall speak for themselves.

Typha latifolia

Typhaceae, Cattail family

OTHER COMMON NAMES:
Cat-o-nine-tail, Flag, Bulrush

CATTAIL

The Cattail is divided into several species which are found in marshes, shallow water, ditches and stagnant pools throughout most of North America. The common Cattail's long, slender leaves and soft, brown, cigar-shaped seedhead hardly need description. They are such a familiar sight not only in the country, but in ditches along freeways, and in flower arrangements in homes and store windows downtown.

It has been said that one could not starve living next to a Cattail marsh. I think this is close to the truth.

Let us start in the spring. The new leaves that shoot up out of the cool water can be pulled up. At their base, where they come out of the rootstock, they have a tender core a few inches long that can be used in a salad or cooked. This is called "cossack asparagus" and those that grow in the marshes of the Don River in Russia are avidly consumed by the natives. Sturtevant's extract from Clarke's *Travel in Russia* indicates that he found the people devouring it raw " 'with a degree of avidity as though it had been a religious observance.' It was to be seen in all the streets and in every house, bound into faggots. They peel off the outer rind and find a tender white part of the stem which for about its length of 18 inches, affords a crisp, cooling and very pleasant article of food." (11)

In early summer the part that later becomes the "cattail" is forming. It is composed of two cigar-shaped spikes one above the other. The top one is the pollen bearing staminate, or male flower, and the one beneath is the seed bearing pistillate, or female flower that later turns into the brown cattail. Just before it comes into bloom both members are green and wrapped in a sheath or husk. At this stage the top staminate spikes can be picked and steamed, or boiled for about five minutes and served like miniature corn on the cob with salt and butter. These are my favorites. They have their own unique flavor which to me is reminiscent of corn (though one source says they are suggestive of both olives and French artichokes). Even if you don't like olives, artichokes, or corn, there still is a good chance you will like Cattail tops.

A little later in the summer the spikes will shed their wraps and the nuptial delights begin. The green staminate spike swells and begins its lusty shedding of yellow pollen which spills down over the pistillate flower spike below and, with the help of a breeze, the pollen is dispersed across the marsh to insure a healthy cross pollination of the rest of the Cattail population. We too can share in the bliss if we are willing to wade out among the blossoms. By inverting and shaking each well endowed spike into a paper bag, a substantial amount of pollen can be collected in very little time. This pollen can be mixed with equal parts of flour and made into wholesome yellow-tinted breads, pancakes, biscuits or pastries, and can be used like saffron to color rice dishes.

The rhizomes produce rapid growing, tender, white runners (as shown on the right side of the drawing). These runners, if caught before they shoot upwards into the light, are good eating peeled in a salad or cooked. The sections of rhizomes between the leaf-clumps have a starchy core that can be utilized as flour. One method of extracting the flour is to dry the roots, grind them up, and sift them to separate the flour from the fibrous cortex. I use the "Gibbons method" which entails washing and peeling the fresh roots. The cores are then put into a bowl of water and crushed by hand until the fibers are all separated. The sediment is allowed to settle for about half an hour and then the water is carefully poured off, leaving the starchy residue on the bottom. More water can be added and subsequently poured off after further settling to more fully refine the product. After the last settling, pour off as much of the liquid as possible, add a little regular flour to thicken it and use it to make your favorite breadstuff.

At the junction of the stalk and the rhizome is a thickened area of starch which, after peeling, can be cooked like a potato and added to stews or soups. Because they are rooted in the oxygen deficient muck of stagnant pools and marshes, Cattails, like many aquatic marsh plants, manifest certain adaptations that allow them to thrive under these conditions. Unlike most land dwelling plants, getting water to the leaves is no problem, but getting oxygen to the roots is something else.

To accomplish this air transfer the leaves have large air vessels or ducts. These can be easily seen by cutting a cross-section at the base of the leaves. The spongy cortex of the rhizome that we peeled off to get at the starchy core serves to pass the air along the rhizome to eventually supply the roots.

I don't know if I have established that it is impossible to starve next to a Cattail patch, but with all the processes I've described, you would not suffer from boredom while you were wasting away.

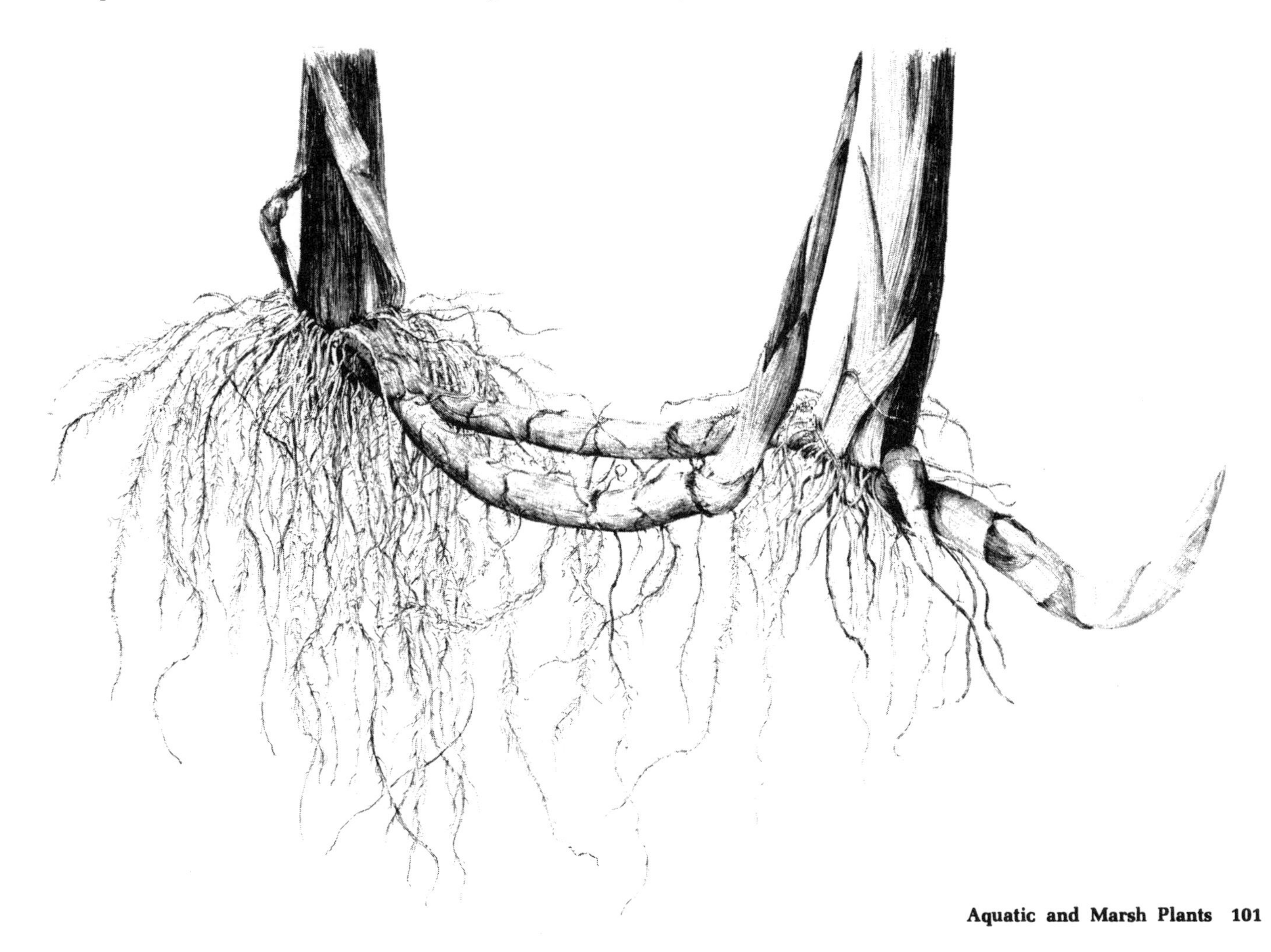

Acorus calamus

Araceae, Arum family

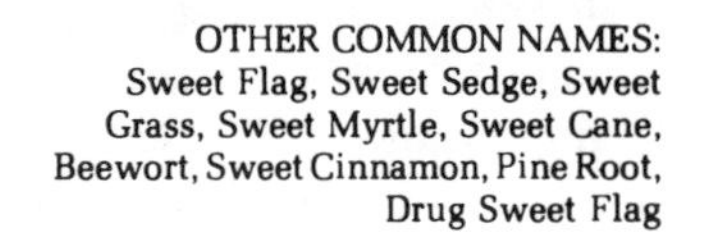

OTHER COMMON NAMES:
Sweet Flag, Sweet Sedge, Sweet Grass, Sweet Myrtle, Sweet Cane, Beewort, Sweet Cinnamon, Pine Root, Drug Sweet Flag

CALAMUS

Calamus is found in swamps, wet meadows and along streams in the east, from Canada south to the Gulf of Mexico and westward, locally, into Montana and Oregon. The leaves are sword shaped and can be three feet long and grow right out of the rhizome. The tender hearts of the leaf clusters in the spring have the spicey Calamus flavor in a mild enough form to make them quite delicious as a nibble or in a salad. You are likely to find them while gathering "cossack asparagus" (see cattail, pg. 100), and the combination of the two, especially with dandelion greens and cress, make an excellent spring salad.

The tiny, yellowish flowers are on a miniature two to four inch spadix which protrudes at an angle about halfway up one of the leaves. Since the fruit and seeds are apparently nonexistant in this hemisphere, the plant propagates by its rapid growing rhizome.

From a distance, Calamus appears to resemble yellow-green cattail shoots or some other marsh grass or sedge. Upon closer investigation, the spicy aroma of the broken leaves and rhizomes confirms its identity. Calamus has a long and active history. It is mentioned in several places in the Bible. Calamus, along with myrrh, cinnamon, cassia and olive oil were the "finest spices" from which Moses was to make a holy annointing oil (Exodus 30: 23-25). Pliny and Dioscorides mention a plant called *Acoron* which, from its description, is presumed to be Calamus. It has been used in the Orient and India for centuries. In India, Calamus was considered such an important medicine for children that an herb-seller faced serious punishment if he refused to open his shop to supply it at any hour of the day or night.

Calamus, in Eurasia, is thought to have been native to marshes in the mountainous areas of India and China, and to slowly have been introduced into the rest of the continent. The Tartars (Mongolians) are credited with bringing Calamus to Russia and Poland during their conquest of these areas in the eleventh century. They believed that Calamus purified drinking water and always planted it when they planned on settling a new territory. Because of this habit, Calamus became a symbol of the Mongol invasion of these countries, where it since has been bitterly referred to as "Tartar" or "Mongolian poison."

Clusius, a famous Austrian botanist, cultivated it in Vienna in 1574 from roots he obtained from Asia Minor. He distributed Calamus to other botanists in western Europe and by 1588 it was recorded growing well in Germany. In England it was introduced in the last part of the sixteenth century, being first cultivated by John Gerard, the famous herbalist. When the first explorers came to the New World it was well established and abundant in North America. It is unknown whether it travelled with stone age nomads across the Alaskan land bridge from Asia or was here before the peopling of this continent.

Calamus was used in Europe as a strewing herb. Peasants gathered arm loads of fresh Calamus reed and spread it on the floors of their cottages. When it was trodden upon, it released its fragrant aroma, which served not only to perfume the air but also to repell insects. Churches in England used Calamus in the same manner on special religious holidays. In the trials of Cardinal Wolsey, during the scandalous era of Henry VIII in England, one of the charges of extravagance brought against the ostentatious cardinal was not his numerous, luxurious dwellings, but his habit of strewing his floors with Calamus rushes. Since it didn't grow near London, the plant had to be brought at considerable expense from Norfolk and Suffolk.

The candied roots were a popular spicy confection and were used as an aromatic bitter tonic to tone and settle the stomach and relieve indigestion and heartburn. It was listed in the United States Pharmacopea from 1820 to 1916, and in the National For-

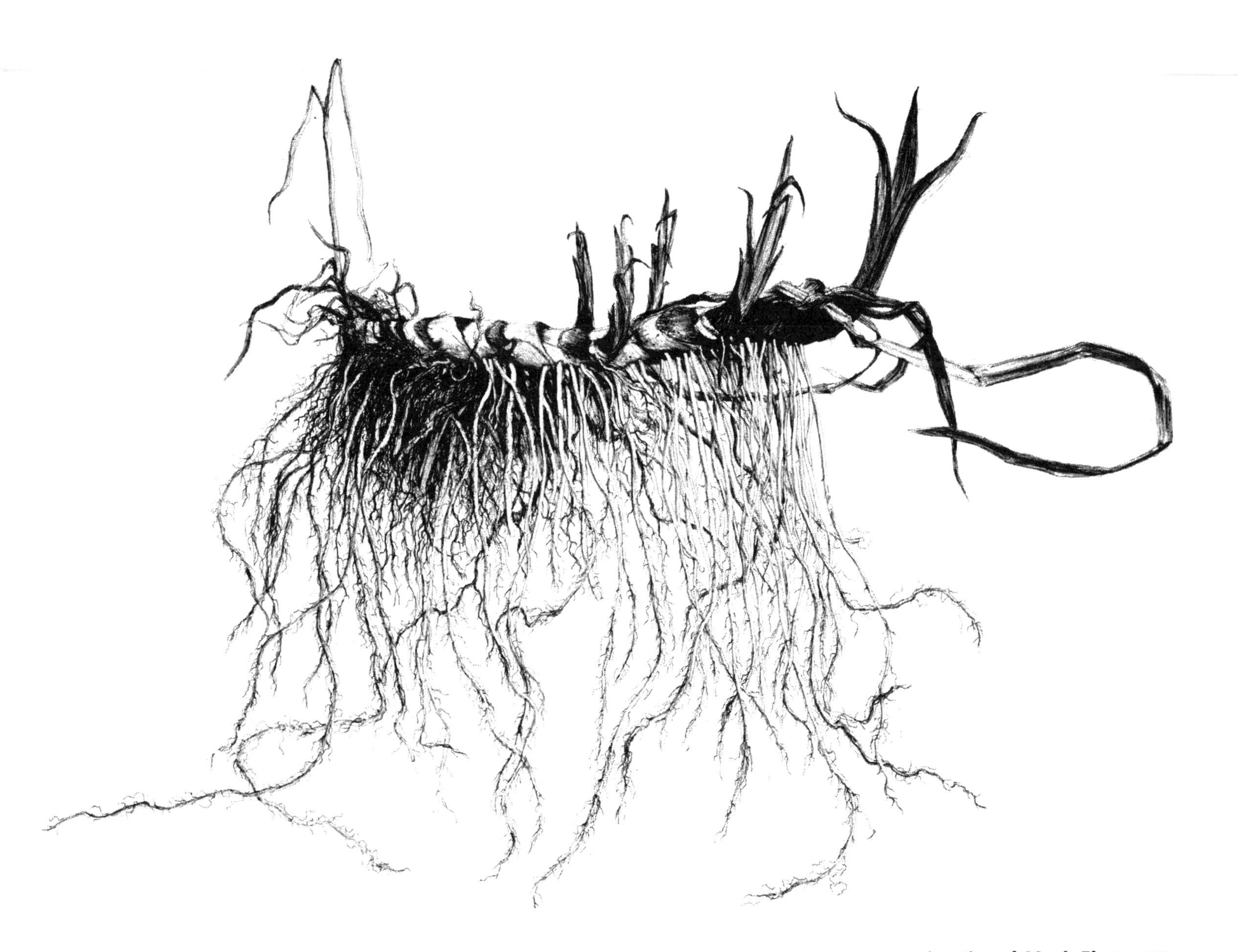

mulary from 1936 to 1950. The tea is used like the candied root, as a stimulant carminative tonic and also externally as a wash applied to sores and burns.

Calamus is said to destroy the taste for tobacco. It is recommended that a piece of the root be carried and chewed whenever the urge for tobacco is felt.

I keep a piece of Calamus root on the dashboard of my car, next to my trusty ginseng root. On long drives I nibble the root, and I find that its pungent flavor and stimulating qualities help me stay awake.

The stimulation apparently comes from a chemical known as *Asarone* which is found in the roots and rhizomes and is chemically related to mescaline and the amphetamines. Reportedly, eating a piece of rhizome two inches long will produce stimulation and "buoyant feelings" and a 10 inch piece will produce hallucinations. Like many natural highs, there is a catch. As soon as you try to eat even a half an inch of the rootstock, you will probably find that it is simply too pungent and strong tasting to swallow.

Indians used Calamus medicinally in many of the same ways as the white man, but the Dakotas had one practice that was different. They chewed the Calamus rootstock into a paste and spread it generously over their faces before going into battle to enable them to remain calm and unafraid in the face of danger. Possibly, the mind altering qualities of the asarone could be absorbed through the skin.

The Calamus rhizomes, when growing in shallow water, form a large entwined mat. The best time to gather them is in spring or fall. They can be cut free and pulled up in great quantities with persistance and a sharp shovel. The rhizomes have rapidly spreading growing tips which creep along the bottom, sending thousands of white hairy roots into the silt and many clusters of leaves skyward. In the illustration of the early winter dug rhizome, the five sprouts that are formed toward the right end of the root are waiting for the warmth of spring. Some of the dead leaves from the previous summer are still draped around. The leaf scars from previous years are clearly visible, particularly toward the left end of the rhizome.

Althaea officinalis

Malvacea, **Mallow family**

OTHER COMMON NAMES:
Sweet Weed, Mortification Root

MARSH MALLOW

Marsh Mallow is a native of Europe, and since its introduction in this country (probably via ship's ballast) it hasn't spread far from the coast. Usually it is found in brackish marshes of tidal estuaries along the northern half of the eastern seaboard.

The plant is two to four feet high and the soft downy leaves are toothed, somewhat resembling a maple leaf in shape. The flowers are one or two inches across, with five pink petals and a protruding pistil surrounded by clusters of stamens.

While few people are familiar with the plant, almost everyone knows the name "marshmallow." The puffy confection we roast over campfires today is actually an offspring (or should I say a corruption) of an old time confectionery paste made from the mucilaginous extract of Marsh Mallow root. This true Marsh Mallow confection was used as a reliable and pleasantly soothing cough and sore throat remedy. Modern marshmallows contain no Mallow.

The Marsh Mallow roots are good eating sliced crosswise, parboiled, drained, and then gently fried in butter or oil with onions and other seasonings. The water in which the roots have been boiled is very viscid and can be beaten like egg whites into a froth. Euell Gibbons describes how he uses them to make such gourmet delights as chiffon pies. (12)

Medicinally, this Mallow water can be taken as a syrup for coughs and sore throats. It can also be used as a hand lotion to sooth chapped or dry skin. This Mallow is also highly recom-

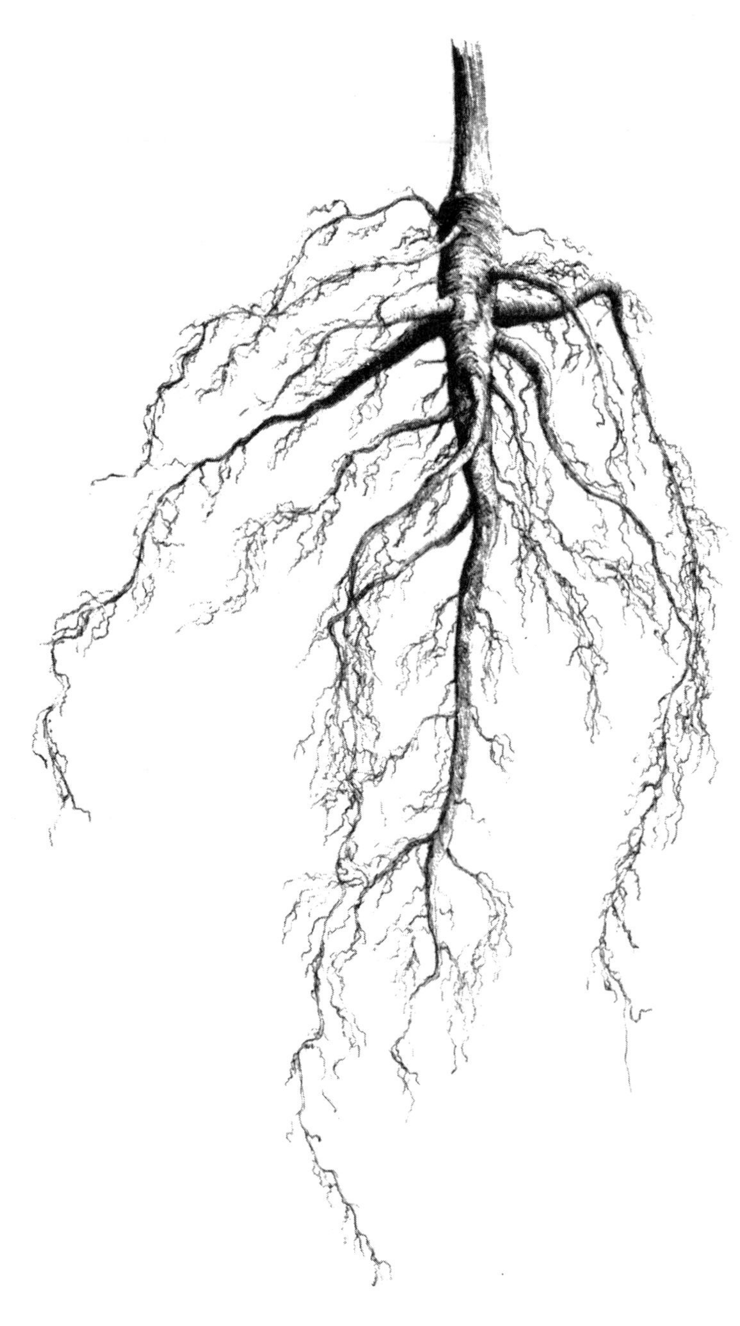

mended when taken internally to ease an ulcerated or abraded stomach or intestinal lining. To preserve it for use over a long period of time add a little alcohol or vinegar.

The roots and upper parts of the plant have a long history of use as a hot poultice applied to draw the infection from wounds and sores, or to soften the skin preparatory to extracting a splinter. Fortunately for those of us who don't live in Marsh Mallow's range, most of the other members of the Mallow family also abound in mucilaginous properties and can be used in much the same way as Marsh Mallow, although it should be mentioned that the inland varieties usually have tough, woody roots.

There is a southern Mallow which grows in salt and brackish marshes from New York to Florida and Louisiana. It is known as *Kosteletzkya virginica*, or Seashore Mallow. Like the other mallows, boiling the root produces the mucilaginous Mallow water. I tried beating the viscid Seashore Mallow solution with a hand egg-beater, attempting to obtain a chiffon-like froth similar to what the true Marsh Mallow produces, but found that my own froth appeared long before there was any response from the Mallow water. However, I sauteed the roots after boiling them, and found them quite good eating.

The leaves and flowers can also be used medicinally in the same fashion as the root. While the upper parts of the plant are best gathered in late summer, the roots can be gathered at any time of year.

Eryngium aquaticum

Umbelliferae, Parsley family

OTHER COMMON NAMES:
(var. yuccifolium), Rattlesnake
Master, Button Snake Root,
Rattlesnake-flag

WATER ERYNGO

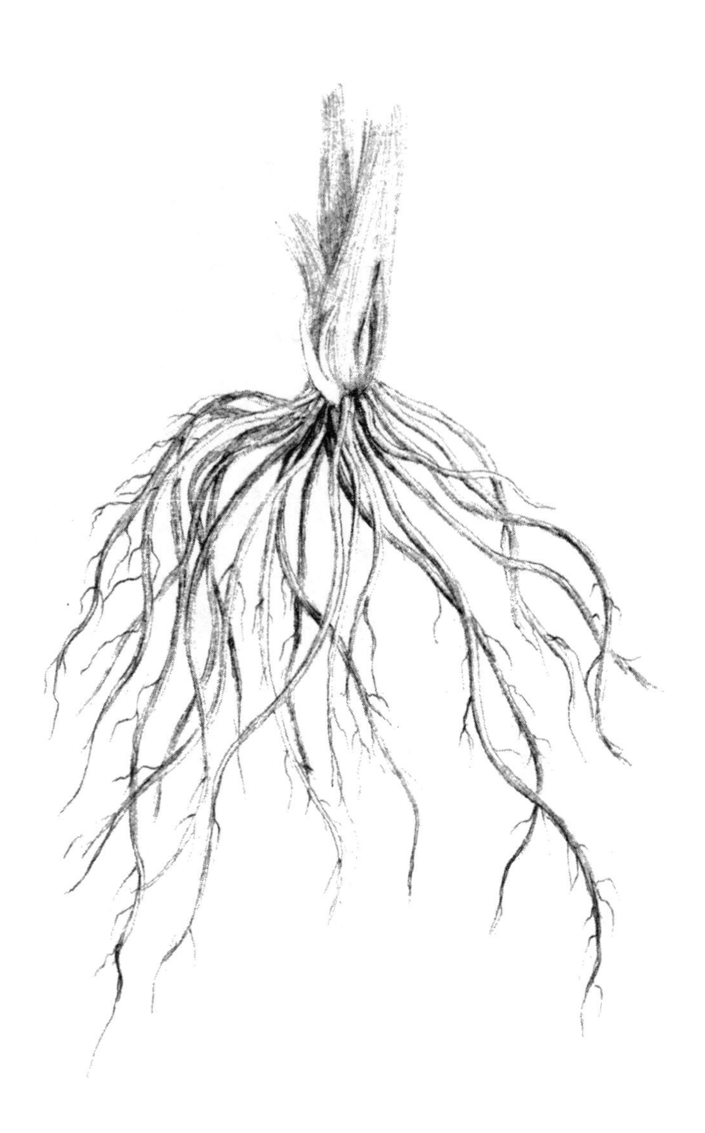

Water Eryngo is found in bogs and swamps and tidewater marshes from the pine barrens of New Jersey south on the coastal plain to Florida and west to Texas. There is also an upland counterpart, *Eryngium yuccifolium,* which is used interchangeably with *E. aquaticum* in herbal practice.

Water Eryngo is two to six feet tall with long narrow leaves that are parallel veined—which is unusual for a dicotyledon—and somewhat resemble yucca leaves. The tiny flowers bloom in late summer in a white, globular head. They often have a blueish cast to them. In fact, I have seen a bog in southern New Jersey where this plant grows so profusely that, during the height of blossoming, the whole area appears as if covered by a blue haze.

Water Eryngo was listed in the United States Pharmacopea from 1820 to 1873. As most of the common names indicate, one of the older uses for the fibrous rootstock of this plant is as a cure for snakebite. It has also been considered useful in the treatment of kidney and liver disorders and is best gathered in autumn.

Limonium carolinianum

Plumbaginaceae, Leadwort family

OTHER COMMON NAMES:
Marsh Rosemary, Canker Root, Ink Root, Marsh Root, American Thrift

SEA LAVENDER

Sea Lavender is common in salt meadows near the ocean on both sides of the Atlantic. There are a few different varieties, but the taxonomists haven't quite come to terms with them all yet. Like most plants that live in salty or dehydrating conditions, the leaves of Sea Lavender are fleshy and succulant. They are oblong and are arranged in a rosette at the base of the plant. The flowering stalk grows up to two feet tall and is very gracefully branched, with rows of tiny, lavender-colored blossoms. They retain their color and make a beautiful addition to dried flower arrangements if picked in the height of bloom.

The thick, reddish, perennial root is the part used in medicine. It is valued primarily as an astringent tonic and is recommended as an externally applied decoction for ulcers and slow healing sores. Sea Lavender was used on the battlefields of Britain in ancient times as a means to stop bleeding.

The plant earned the name canker root because of its use in healing canker sores. As with most astringents, it is valued in treating the symptoms of dysentery and sore throat.

Sea Lavender root was official in the United States Pharmacopea from 1830 to 1870. It can be gathered at any time of the year.

Nuphar advena

Nymphaeceae, Water Lily family

OTHER COMMON NAMES
Yellow Pond Lily, Bonnets,
Beaverroot, Cow, Frog, Dog, Horse,
Bullhead, or Beaver Lily

SPATTERDOCK

Spatterdock is found throughout most of the United States east of the Rockies. There are several varieties, which some sources call races and others call species. They are found in pools, shallow ponds, slow moving streams, and at the edges of tidewater.

The leaves are roundish, with a V-shaped slit where the leaf stem joins. Unlike the white water lilies, the leaves stand upright above the surface of the water rather than floating upon it. The blossom consists of a cup-shaped cluster of fleshy, yellow, petal-like sepals. Inside this cup are numerous, inconspicuous petals.

The western species of Spatterdock produced seeds particularly valued as food by the Klamath tribe. These were dried and then cracked from their shells. The kernels, once extracted, were parched and eaten like popcorn, or ground into flour to use for making breadstuffs. In the east, Indians reportedly ate the thick, creeping rootstock. The squaws dove for it, or raided muskrat houses where the animals often stored a supply for themselves. A comparable "trade" of some other food was customarily made, so as not to deprive the hungry muskrat clan.

Many of the wild-food books say that the rootstock can be boiled, baked or roasted. When I have tried this, and no matter how I cooked it, the taste was of boiled, baked or roasted SWAMP! I wonder how much personal experience is involved in these modern reports. There probably is a method of making the root palatable, but so far I have not been able to find it.

Whether it is palatable or not the rhizome is a very impressive sight. It is up to five inches in diameter and has a very light colored flesh. The upper surface is covered by uniform, dark leaf scars. And the myriad roots extend from the underside, anchoring it into the mud. The rhizome is perennial and continues growing year after year. I have never been able to measure the entire length of one, but it is interesting to speculate that one rootstock might extend completely across a whole pond, regularly branching and sending up leaves and flowers. In fact a whole pond full of Spatterdock could actually be just one plant!

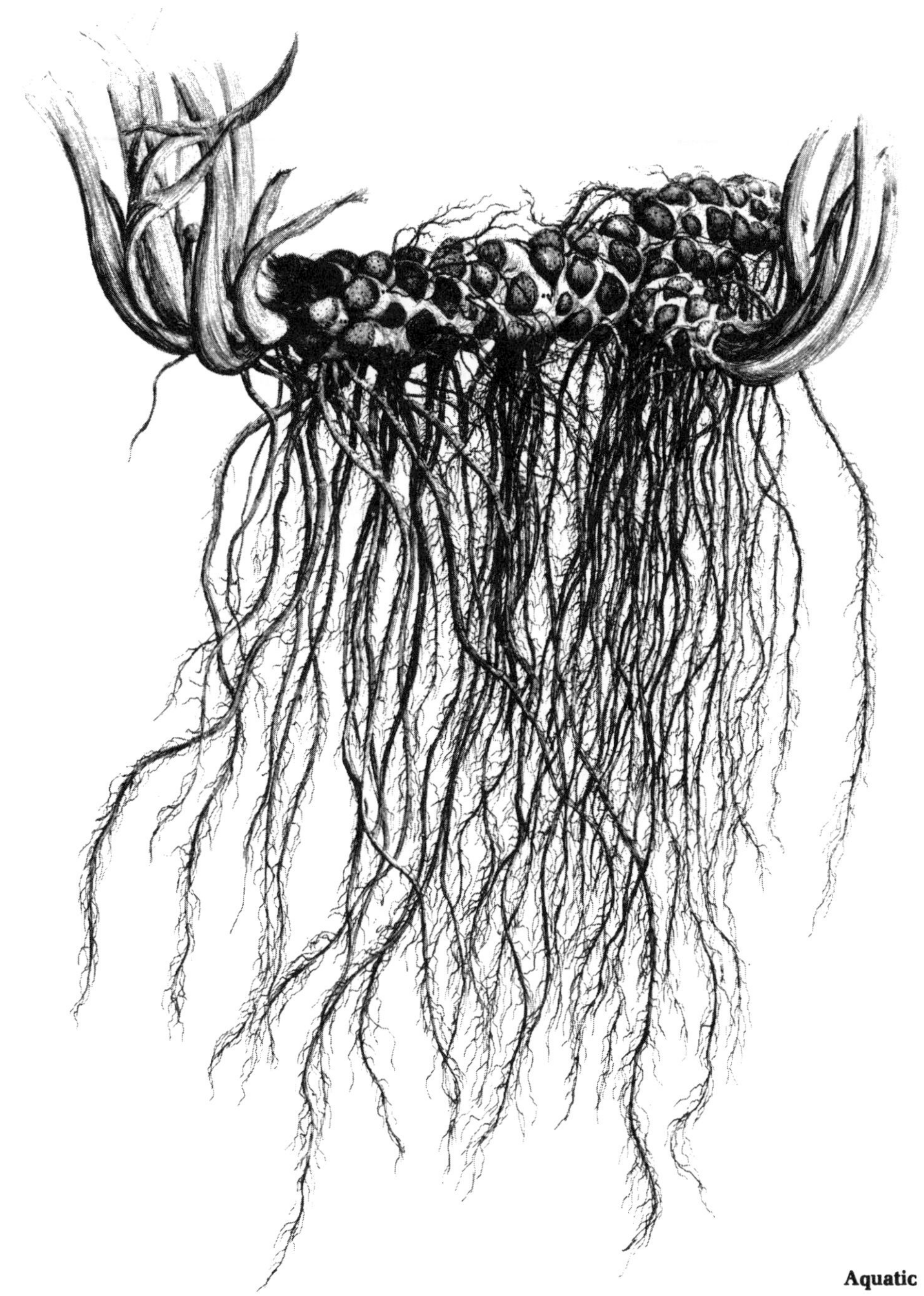

Symplocarpus foetidus

Araceae, Arum family

OTHER COMMON NAMES:
Skunk-weed, Clumpfoot Cabbage, Polecat-weed, Swamp Cabbage

SKUNK CABBAGE

Skunk Cabbage is found in swampy, wet woods and in shaded areas along streams east of the prairies, from Canada south to Iowa and the Carolinas. Like its name suggests, Skunk Cabbage resembles a large spreading head of cabbage, and the smell of the bruised leaves recalls a skunk.

Skunk Cabbage is well known for its early spring appearance. Often before the snow has melted, the pointed tips of the Skunk Cabbage can be seen protruding from the icy muck, like tightly sealed, fleshy tepees. It is surprising to know that in order to make this early appearance, these spikes first form late in the summer before, and remain dormant over most of the winter. Sometime in February or March these little tepees begin to churn with growth activity. This internal activity is furious and generates amazing amounts of heat. It has been recorded that the temperature inside these feverish tips can be 27° F. warmer than the winter air surrounding them. This allows the Skunk Cabbage to withstand the winter freezes, and to create its own springtime within. As we get to the spring side of winter, the teepee, which like the Jack's pulpit is really called a spathe, swells, and its color brightens to a yellow or spotted maroon. Its small, slit-like door opens wide enough to reveal a knobby spadix which is covered by the bright yellow anthers of the blossoms. These form into seeds that are imbedded in the spadix. Long after the great summer leaves have died, the spherical spadix can often be found lying in the ooze like a child's long forgotten ball. The tubular cluster of leaves that follow the spadix blossom, if caught early in the year, can be eaten when boiled in several waters. Some wild-food fanciers claim that they are quite palatable. I have eaten them and found that although they were fairly good tasting, they have a bit of an unpleasant bite from the calcium oxalate acid crystals they contain. However, both the leaves and the roots can be prepared as was described in the chapter on Jack-in-the-pulpit by exposing them to air and dryness for a long period of time.

Reportedly, different populations vary greatly in their edibility. The forager should also be careful not to mistake hellebore (*Veratrum viride*) for Skunk Cabbage. The leaves of hellebore, growing near or among Skunk Cabbage, are much more deeply and sharply veined, and it is quite poisonous.

Skunk Cabbage has become valued by humanity more for its medicinal virtues than its food value. The rhizomes and roots were listed in the United States Pharmacopea from 1820 to 1882. It is classed as an emetic, diuretic, stimulant and narcotic, but it is most renowned for its antispasmodic properties for which it has earned a time-honored place in the form of a tincture that is highly valued by herbal practitioners. This tincture is used to treat all ailments that involve spasms, cramps or constrictions of any kind, including spasmodic coughs, asthma, cramps, epilepsy, lockjaw, delirium tremens, rheumatism, etc.

To make an antispasmodic tincture, you need one ounce each of Skunk Cabbage root, black cohosh root, lobelia seed and myrrh gum, with one half ounce of cayenne. Some recipes also include valerian root and lobelia herb. From these ingredients, either an alcoholic or acetic tincture can be made. To make the alcoholic tincture, the ingredients are crushed, combined, and steeped for one to two weeks in a pint of pure grain alcohol, and shaken daily. Subsequently it is filtered and the resulting liquid is bottled for use. To make the acetic tincture, the herbs are boiled in one pint of water for one-half hour and, after straining, the resulting liquid is added to one pint of apple cider vinegar and bottled for use.

The resulting tincture may be used externally for cramps, swellings and rheumatism. Internally, it is administered in doses of usually less than a teaspoonful in a cup of tea or water.

An ointment made of Skunk Cabbage rootstock is used to treat various external ailments such as ringworm, sores, and painful swellings.

The thick rhizome of Skunk Cabbage penetrates deep into the ground and is enclosed by whorls of smaller roots. It involves a fair amount of shoveling to unearth but the rhizome is large enough that one is usually sufficient for the average person's needs.

The root loses its potency fairly rapidly, so it is recommended that you renew your supply at least once a year.

Nymphaea odorata

Nymphaeaceae, Water Lily family

OTHER COMMON NAMES:
Sweet-scented White Water-Lily,
White Pond Lily, Water Nymph,
Water Cabbage

FRAGRANT WATER LILY

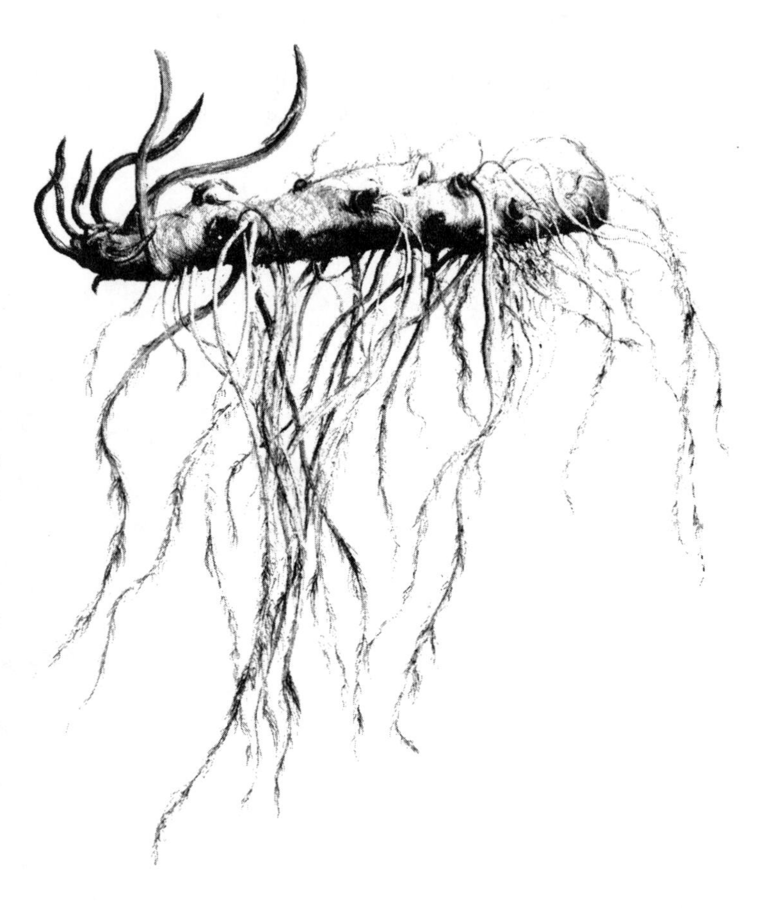

The Fragrant Water Lily is found in slow streams, lakes and ponds east of the prairies, from Canada south to the Gulf of Mexico. This is the common plant with a white, fragrant blossom and the round floating "pads" so familiar to many of us.

Both the unopened flower buds, and the unfurling leaves, can be picked and eaten after a few minutes of boiling.

The rootstock was reportedly used for food by Indians. Medicinally, the rhizome and roots are classed as astringent, antiseptic and demulcent. A decoction of the root is used to treat diarrhea, as a gargle for sore throat, and a wash for sore eyes. As a douche, it is highly recommended against vaginal infections. The powdered root is used as a poultice for sores and skin irritations, where it is recommended that it be combined with slippery elm or crushed flax seeds.

The perennial rhizome has numerous scars from the leaves. When it is broken and exposed to air, it takes on a purplish tinge. If it is cut in cross-section, a circle of large, porous air chambers can be seen. The rhizome is covered with fine hairs which, upon drying, give it a silvery, silky appearance. The rhizomes can be gathered at any time of the year.

Cicuta maculata

Umbelliferae, Parsley family

OTHER COMMON NAMES:
Beaver Poison, Spotted Cowbane,
Musquash Root, Poison Parsley,
Children's-bane, Spotted Hemlock,
Carotte a moreau (French Canadian)

WATER HEMLOCK

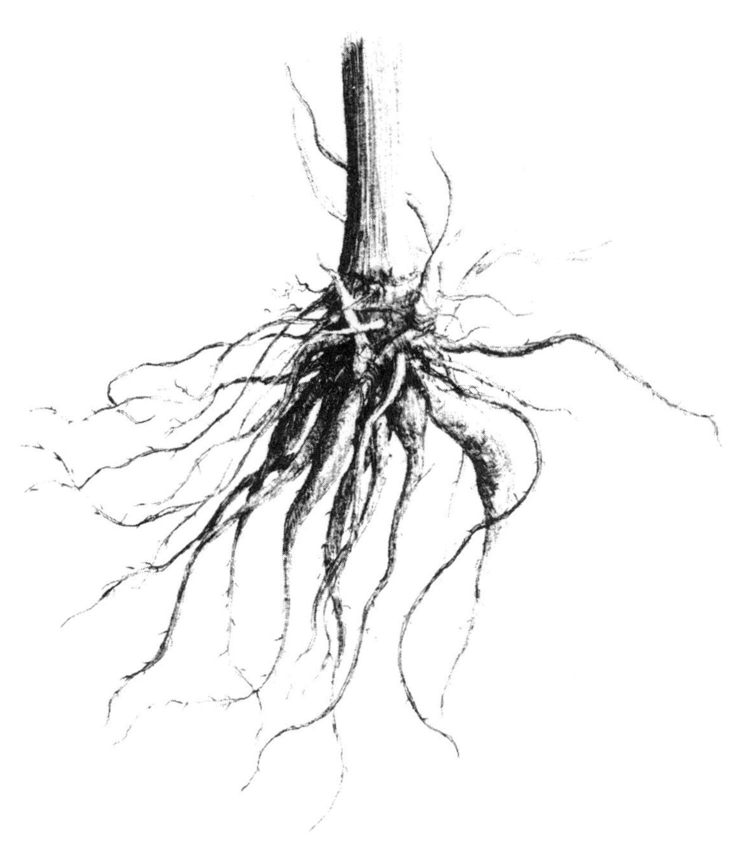

Water Hemlock, as its name indicates, is found in swamps, along streams, and in low ground from Canada, east of the prairies south to Florida and New Mexico.

Neither the plant nor the root has any beneficial use. Its inclusion here is simply to provide an example of a plant whose root might appear to be attractive, but which is, in fact, deadly.

Its reputation as our most virulent poisonous plant comes more from the violence of the symptoms it produces than from the strength of its toxin. It is a severe convulsant, as violent as its cousin the poison hemlock (Conium) is gentle.

It is a perennial plant, three to eight feet tall, with a purple streaked stem. The compound leaves may be more than one foot long. The individual leaflets are elliptical, toothed and usually one or two inches long. An outstanding field mark of this Hemlock is that the secondary lateral veins in these leaflets, instead of going to the points on the teeth, end in the notches between the teeth. The plant blossoms in early summer. The flowers are in an umbel three to five inches in diameter and seeds are produced in autumn.

The thickened portions of these roots serve as food storage reservoirs. The food in each cluster of roots is exhausted in the production of the flowering stem during the summer. The foliage, however, lasts for the entire growing season and builds up a new cluster of thickened roots which remains in the ground over the winter when the original plant dies off. In the spring it produces new foliage and a new flowering stalk. In this way a population is able to remain in the same place from year to year and, of course, it is able to spread by the seeds.

The roots are the most toxic portion of the plant—as little as a mouthful can kill a person. Many poisonings occur in the spring when the roots are exposed by spring rains and flooding, and are mistaken for parsnips or other edible roots. I have pulled them up on occasion, and have to agree that they do have a very pleasant odor resembling a freshly cut carrot or parsnip. In fact, the first time I found them, they smelled so good to me that if I didn't adhere strictly to a policy of not tasting a plant until I was sure of its identity, I might not be writing this today. According to the reports, the poisoning victims have said that the roots not only smell good but they taste good too.

This graphic description of a case of Water Hemlock poisoning in Europe was originally written in Latin in 1679. It should be sufficient to discourage wild food fanciers from eating plants they can't positively identify.

> When about the end of March, 1670 the cattle were being led from the village to water at the spring, in treading the river banks they exposed the roots of this *Cicuta* (water hemlock), whose stems and leaf buds were now coming forth. At that time two boys and six girls, a little before noon, ran out to the spring and the meadow through which the river flows, and seeing a root and thinking that it was a golden parsnip, not through the bidding of any evil appetite, but at the behest of wayward frolicsomeness, ate greedily of it, and certain of the girls among them commended the root to the others for its sweetness and pleasantness, wherefore the boys, especially, ate quite abundantly of it and joyfully hastened home; and one of the girls tearfully complained to her mother she had been supplied too meagerly by her comrades, with the root.
>
> Jacob Maeder, a boy of six years, possessed of white locks, and delicate though active, returned home happy and smiling, as if things had gone well. A little while afterwards he complained of pain in his abdomen, and, scarcely uttering a word, fell prostrate on the ground, and urinated with great violence to the heighth of a man. Presently he was a terrible sight to see, being seized with convulsions, with the loss of all his senses. His mouth was shut most tightly so that it could not be opened by any means. He grated his teeth; he twisted his eyes about strangely and blood flowed from his ears. In the region of his abdomen a certain swollen body of the size of a man's fist struck the hand of the afflicted father with the greatest force, particularly in the neighborhood of the ensiform cartilage. He frequently hiccupped; at times he seemed to be about to vomit, but he could force nothing from his mouth, which was most tightly closed. He tossed his limbs about marvelously and twisted them; frequently his head was drawn backward and his whole back was curved in the form of a bow, so that a small child could have crept beneath him in the space between his back and the bed without touching him. When the convulsions ceased momentarily, he implored the assistance of his mother. Presently, when they returned with equal violence, he could be aroused by no pinching, by no talking, or by no other means, until his strength failed and he grew pale; and when a hand was placed on his breast he breathed his last. These symptoms continued scarcely beyond a half hour. After his death, his abdomen and face swelled without lividness except that a little was noticeable about the eyes. From the mouth of the corpse even to the hour of his burial green froth flowed very abundantly, and although it was wiped away frequently by his grieving father, nevertheless new froth soon took its place. (13)

IV
OTHER ROOTS

There are numerous roots that I have chosen not to illustrate or discuss due to the relative scarcity of the plant, or the plant's lack of a particularly unique root form. Since many of these roots are, none the less, valuable parts of our underground flora, I will at least mention some more of them in passing.

Many people know the Bayberry *(Myrica cerifera)* for its wax-producing berries and fragrant leaves. Though not the same as the bay leaves of commerce, Bayberry leaves can be similarly used to flavor food and make tea. The bark of the root, however, has long been known as an astringent and a stimulant, used as a gargle for sore throat, as a mouth wash for sore gums, as a tea for diarrhea and jaundice, as a douche for uterine problems, and in its powdered form as an application to slow healing sores and boils.

The common Blackberry *(Rubus spp.)*, which we all know for its fruit, has a high tannin content especially in its root and it is used in a decoction to treat all forms of diarrhea and dysentery.

Yellow Dock *(Rumex crispus)*, as its name indicates, has a yellowish root which is regarded as gently laxative, alterative and tonic. It is used to tone the system, purify the blood, and is also made into an ointment for skin problems.

The root of the tiny Goldthread *(Coptis groenlandica)* is also well named because it resembles a golden thread that winds its way through the mire in moist areas beneath firs and hemlocks of the North. The root is as bitter as it is bright and can be used as an excellent bitter tonic to tone the stomach, or chewed to cure canker sores.

The Virginia Snakeroot *(Aristolochia serpentaria)*, also known as Serpentaria, was listed in the United States Pharmacopea as a diaphoretic, aromatic, bitter tonic and stimulant used for treating lung diseases, fevers and stomach disorders.

Another root, called the Seneca Snakeroot *(Polygala senega)*, is a little more severe in action than its previously mentioned name-sake. It was used as expectorant, diuretic and diaphoretic but can be a severe irritant in large doses. It was used mainly to treat various lung problems, from asthma to bronchitis and pneumonia. It was an official drug in the United States Pharmacopea and the National Formulary from 1820 until 1960.

The root of a small plant with bright red, tubular flowers, called Pinkroot *(Spigelia marilandica)*, has been used to expell intestinal worms.

Culver's root *(Veronicastrum virginicum)* is another medicinal root that was found in commerce until recently. It was used as a general tonic, strong laxative, and to treat liver problems.

The Blue Flag *(Iris versicolor)* is a handsome, swamp-dwelling relative of the garden iris. The perennial rhizome has been powdered and added in small amounts to blood purifying compounds. It is a purgative and is used as a hepatic stimulant in liver problems.

Growing in the swamps, sometimes near skunk cabbage, is the Green Hellebore *(Veratrum viride)*. The pleated leaves are sometimes mistaken for skunk cabbage—with drastic results; for the whole plant, especially the root, contains powerful drug compounds, most notably *protoveratrine*, which in miniscule doses lowers blood pressure. It is sometimes used in contemporary medical practice in combination with *Rauwolfia*, or Indian Snakeroot, to treat high blood pressure and heart problems. Because the exact dosage is so critical, this is not an herb to be used as a home remedy.

One of the favorite foods of American Indians was the tuber of the Wapato or Water Nut *(Sagittaria latifolia)*. This is a pond and swamp dwelling plant with arrowhead shaped leaves. In the late fall, after the leaves have died off, the plant produces edible white tubers that can be cooked and eaten like potatoes. Indian squaws would wade into the cold water and dislodge the tubers with their toes. If the November chill is in the water, a more comfortable alternative to wading is to rake the muddy bottom with a potato rake. The tubers will float to the surface.

Another useful tuber is that of the Chufa or Nut Grass *(Cyperus esculentus)*. One of the few tuber-bearing, edible sedges, the Chufa is fairly common, especially in the south; but because of the similarity among sedges, it is not very conspicuous. In some parts of the deep south, Chufas are cultivated and, because they are nut-like, are sold under the name "Florida Almonds."

The first year taproot of the biennial Evening Primrose *(Oenothera biennis)* is edible. It must be harvested at just the right time in the spring or it will have an unpleasant aftertaste. It is, however, cultivated for food in Germany today, and called German rampion.

One of the first American food roots to be written about in English was the acrid Virginia Tuckahoe *(Peltandra virginica)*. John Smith wrote that the Virginia Indians roasted them in pits, no doubt to disperse the characteristic arum acridity.

The Seminoles of South Florida still harvest the large, starch-bearing Coontie Root of a small native cycad *(Zamia floridana)* which they use in making breadstuffs.

There is another plant of the coastal savanna plain of the South called Redroot *(Lachnanthes tinctoria)* which does have a red root that can be used in dying wool.

Many people know the Cotton plant *(Gossypium spp.)* for its blossom and the fibers that enclose the seeds. It is less well known, however, that the Cotton root contains a substance that increases the contractions of the uterus during childbirth. Also, the root has a long tradition of use as an abortifacient.

The woodland shrub *Euonymus*, known popularly as Wahoo, Spindle Tree, Bush Strawberry, or Heart's-a-bustin' is most conspicuous in late fall, because of its bright crimson and magenta berries. Its root has often been used as a digestive and liver tonic. In order to "uncross" someone who has been the target of a witch's evil spell, a piece of the root is held over the victim's head and the word "wahoo" is repeated seven times.

Appendix:

Harvesting Roots

Sooner or later you may want to gather some roots for your own use. Obviously some digging tools are necessary. These can be as sophisticated as the post-hole digger described in the section on burdock, or as simple as a "sengin" stick used by ginseng diggers in the Appalachians. Because ginseng and many other forest roots grow in the moist, soft soil of the rich hollows, it takes little effort to dig them once they are located. A small trowel or digging stick are all that are necessary. To make a "sengin" stick, a "seng digger" cuts a sapling about the diameter of a broom handle and whittles a chisel-shaped point. I usually like to find a hardwood sapling with a smaller branch or stub six or eight inches from the digging end. This serves as a footrest so that the stick can, if necessary, be pushed into the ground like a shovel. If the stick is cut to be four or more feet long, it can double as a hiking staff. To make it a real work of art, you can harden the point by heating it in the coals of the campfire, tie feathers to it and carve sacred totem symbols in the bark. Of course, then you will have to worry about losing it. I have tried to establish the habit of carrying a small trowel with me, but because I tend to become so enthralled by whatever I have unearthed, I almost always leave the trowel (or digging stick) behind. After littering the woods with far too many trowels, and destroying too many saplings to make digging sticks, I have begun to tie my trowel to my pack basket with a length of twine. Still, after I have dug some roots, carefully packed them in my pack basket and am headed on my way, I am often surprised to hear the trusty trowel clanging along on its leash only a few feet behind.

In fields and prairies the soil is often harder and dryer and the roots, in their search for water, must grow to much greater depths. This is when a bigger shovel or posthole digger might be necessary. I prefer a fairly narrow, sharp pointed spade.

In swamps and shallow ponds the rootstocks can sometimes be cut with a shovel and pulled out of the surrounding silt. Once you have gathered up a supply of roots, it is best to wash them well, while they are fresh. If available, there's no better place to wash them than in a clean, swift flowing brook. If it is a warm day, I usually take the opportunity to wash myself too (downstream from the roots, of course). A garden hose works well also, especially because the high pressure stream can more effectively clean all the cracks and crevices in the root. A small vegetable brush is very handy for scrubbing especially persistant dirt or clay. After the roots are washed, they must be dried. This is best done by spreading them out in a warm, airy place. The flowers, leaves and other green parts of herbs will be ruined if they are dried in direct sunlight, but most roots may be sun-dried. They should, however, be covered or taken in at night. The larger, fleshier roots dry best if they are chopped or split into small pieces. Large sassafras roots are most easily chopped into usable pieces after the roots have been drying for a few days, but before they have turned completely hard. Roots must be absolutely dry before they are sealed in containers, or they will mold — which will destroy their usefulness. I recommend that, until you are absolutely sure your roots are dry, they be stored in paper bags near artificial heat for a month or two, to allow every bit of their moisture to escape.

Since the entire plant is destroyed in gathering most kinds of roots and rhizomes, a special conscientiousness must be exhibited when gathering them. The principle rule is: difuse your impact. Take plants that are growing closely together. Dig roots from the center of a bed so that they will be able to fill in easily. Whenever possible, dig roots after plants have gone to seed in the fall. Rhizomes of some plants, like goldenseal and Solomon's seal, grow horizontally; if these are broken and the portion with the stem and next year's bud is replanted, the plant will usually survive.

These suggestions are just common sense which, coupled with care and consideration is all it takes to use and, at the same time, preserve these valuable subterranean resources.

A Method of Detailed Examination of Root Systems

To accurately study the spread and distribution of a root system in soil, more sophisticated devices than a "sengin stick" are necessary.

One method uses a pin-board. The pin-board is a rectangular board large enough to cover the root system being studied. Long stiff "pins" should protrude in a grid pattern, five centimeters apart. Steel knitting needles are recommended.

When the plant to be studied is located, a pit—the length and breadth of the pin-board—is dug next to it. The wall of the pit adjacent to the plant must be scraped until it is smooth and flat. The pin-board is placed beside the plant and pushed into the wall of the soil. Sometimes a screw-jack braced against the other wall of the pit is necessary to push the pins in. If holes are drilled in the board but the pins aren't inserted until the board is in place, the pins can be pushed or hammered in individually.

With the pin-board in position, a horizontal trench is dug beneath the board, extending a few inches beyond the length of the pins. With the board still supported by the jack, vertical trenches are also dug on either side of the board, to extend a few inches wider than the length of the pins. A thin cable with a handle on each end is placed in the trench and drawn upward with a sawing motion, just beyond the pins. This frees the block of soil which can then be maneuvered into the hole so that the pins are pointing up.

With the help of several people, and/or a hoist, the block is taken out of the ground to the laboratory where it is gently agitated in a water bath to remove the soil.

Once the soil is removed, the board is photographed. The pins, spaced every five centimeters, supply a measureable scale, and they support the root and its branches in very nearly the same position they held in the soil.

Glossary

ALTERATIVE. A substance that alters a condition and gradually restores health (vague and archaic).

ANNUAL. A plant that completes its life cycle and dies entirely in one year.

ANTHELMINTIC. An agent used to destroy or expell parasitic intestinal worms.

ANTIPERIODIC. An agent that cures periodically recurring symptoms of a disease. Malaria treatments are often considered antiperiodic.

ANTISCORBUTIC. A food or medicine that can precent scurvy. Anything that contains vitamin C is antiscorbutic.

ANTISEPTIC. A substance that inhibits the growth of infection-causing microorganisms.

ANTISPASMODIC. An agent that will relax spasms (involuntary contractions) of all kinds. Antispasmodics are used for treating cramps of all kinds, epilepsy, asthmatic attacks, spasmodic coughs and forms of paralysis.

APERIENT. (see laxative)

APHTHOUS. Relating to thrush, or foot-and-mouth disease.

AROMATIC. Plants with a strong smell or taste usually produced by volatile oils (oils that evaporate readily).

ASTRINGENT. A substance that causes tissue to draw together, contract or pucker, and causing a stoppage of mucus or fluid discharge.

ATONIC. Having insufficient muscle tone.

BIENNIAL. A plant that lives for two seasons before it produces seeds and dies.

CALMATIVE. An agent that has a gentle sedative action.

CARMINATIVE. An agent used to relieve gas pains, colic, flatulence or griping, or to expell gas from the intestinal tract.

CATHARTIC. (see laxative)

HYDRAGOGUE. Causing a discharge of water.

CORTEX. The main outer tissue of a stem or root between the outer epidermis and the central core (containing the sap vessels commonly enclosed by an inner wall called the endodermis).

COUNTERIRRITANT. An agent that produces superficial pain that more or less ties up the nerves so a deeper pain won't be so noticeable. The use of counter irritants doesn't represent a high point in the healing arts.

DECOCTION. A boiled herbal tea which releases the bitter or medicinal properties of the herb. Sometimes refers to a more concentrated tea, accomplished by boiling it down.

DEOBSTRUENT. A medicine that clears obstructions from the ducts of the body.

DEMULCENT. A substance that is soothing to the mucous membranes.

DIAPHORETIC. A substance that increases perspiration. This aids in eliminating wastes from the body. Some diaphoretics work by producing heat and are used for colds and chills. Diaphoretics that work without producing heat are used to break fevers.

DIURETIC. Increases the flow and volume of urine, thereby improving or increasing the elimination of waste products through the urine.

EMETIC. Causes vomiting.

EMMENAGOGUE. An agent that stimulates menstrual flow.

EMOLLIENT. The external counterpart of a demulcent. It softens and soothes the skin.

EXPECTORANT. A substance that facilitates expulsion of mucus or phlegm from the respiratory tract.

GLUTINOUS. Having the viscous, sticky properties of glue.

INFUSION. A non-boiled tea where the herbs are steeped in cold or hot but not boiling water. Infusions are used when the aromatic qualities of an herb are desired. The container in which a hot infusion is made should be covered.

LAXATIVE. An agent that increases bowel action. In herbal medicine there are various words that describe the relative strengths of laxatives. From gentlest to strongest, they are as follows: aperient, laxative, cathartic and purgative. The last two especially are not recommended.

MORDANT. A substance used to fix the colors of dies.

MUCILAGINOUS. Gummy, sticky or slimy.

NERVINE. A substance that acts to tone and soothe nerves.

PALMATELY. Having leaflets radiating out from a central point.

PANICLE. A diversely branched, loose flower structure.

PARTURIENT. Relating to the process of childbirth.

PLEURISY. Inflamation of the membrane that envelopes the lung cavity.

PISTIL. The seed bearing organ of a flower.

POULTICE. A moist soft mass of herbs or other substances applied externally to sores, bruises or other affected areas.

PURGATIVE. (see laxative)

RACEME. An elongated flower cluster consisting of a central stem and many flowers each mounted on its own stem. Usually the lower flowers on a raceme bloom first.

SEPAL. The green, petal-like segments encasing the outside of a flower, where it joins the stalk.

SPADIX. A flower cluster on a fleshy central stem.

SPATHE. A large, often showy, leaf-like bract that springs from beneath and often encloses a flower cluster which is commonly a spadix.

SPIKE. An elongated flower cluster where the individual flowers are stemless and are attached directly to the central stem.

STAMEN. The pollen producing organ of a flower, looking like a filament with a tiny knob on the end (anther).

STIMULANT. In herbal practice stimulants increase the various functional activities of the body. They are usually more healthful and general than nerve stimulants like nicotine and caffeine.

STYPTIC. An astringent substance which may, in some cases, stop bleeding.

TINCTURE. An alcoholic solution of medicinal substances.

TONIC. A substance that imparts tone, strength, balance and resiliency, i.e.health, to the body and its systems, organs or tissues.

UMBEL. A flower cluster in which the flower stalks spring from a common center and form a flat or slighly dome shaped surface.

Notes

DILUTION AND EVOLUTION

1. C.M. Wilson, *Roots: Miracles Below,* page 2.
2. E. Epstein, "*Roots*" *Scientific American,* Volume 228 #5, page 48.
3. Howard J. Dittmer, "A Quantitative Study of the Roots and Root Hairs of a Winter Rye Plant *(Secate Cercale),*" pages 417-419.
4. P. Yogananda, *Autobiography of a Yogi,* pages 533-539.
5. Ibid., page 422.

GINSENG

6. Louise Veninga, *The Ginseng Book,* page 51.

KUDZU

7. Fernald, Kinsey & Rollins, *Edible Wild Plants of Eastern North America,* page 258.

POISON HEMLOCK

8. From Plato: PHAEDO, translated by F.J. Church, copyright © 1951, by The Liberal Arts Press, Inc., reprinted by permission of the publishers, The Bobbs-Merrill Company, Inc. Pages 72-74.

SASSAFRAS

9. V.J. Vogel, *American Indian Medicine,* page 362.

JERUSALEM ARTICHOKE

10. Euell Gibbons, *Stalking the Wild Asparagus,* page 27.

CATTAIL

11. Fernald, Kinsey & Rollins, *Edible Wild Plants of Eastern North America,* pages 83-84.

MARSH MALLOW

12. Euell Gibbons, *Stalking the Healthful Herbs,* pages 194-195.

WATER HEMLOCK

13. C.A. Jacobson, "Water Hemlock *(Cicuta).*" Nevada Agricultural Experiment Station, *Technical Bulletin 81,* 1915.

Bibliography

Anderson, Edgar. *Plants, Man and Life.* Berkeley, California: University of California Press, 1971.

Bergen, Joseph, and Davis, Bradley M. *Principles of Botany.* Boston: Ginn, 1906.

Bowen, G.D. and Rovira, A. "Influences of Mirco Organisms on Growth and Metabolism of Plant Roots." *Root Growth.* W.J. Whittington, Ed., London: Butterworths, 1968.

Britton, Nathanial Lord, and Brown, Addison. *An Illustrated Flora of the Northern United States and Canada.* Volume 3, New York: Dover, 1970.

Clymer, R. Swineburne. *Nature's Healing Agents.* 5th Edition, Quakertown, Pennsylvania: Humanitarian Society, 1973.

Crowhurst, Adrienne. *The Weed Cookbook.* New York: Lancer, 1972.

Culbreth, David, M.R. Phg., MD. *A Manual of Materia Medica and Pharmacology.* 7th Edition, Philadelphia, Pennsylvania: Lea and Febriger, 1927.

Culpeper, Nicholas. *English Physician and Complete Herbal.* Arranged for use as a First Aid Herbal, by Leyel, C.F. No. Hollywood, California: Wilshire Book Co., 1972.

Dittmer, Howard. "A Quantitative Study of the Roots and Root Hairs of a Winter Rye Plant *(Secale Cereale).*" *American Journal of Botany.* Volume 24 #7. July 1937.

Elliott, Douglas. "Conscientious Herb Gathering" *Mother Earth News* #28. Hendersonville, North Carolina: July, 1974.

Epstein, Emanual. "Roots." *Scientific American. Vol. 228* #5, May, 1973.

Fernald, Merrit L. *Gray's Manual of Botany.* 8th Edition, New York: American Book Co., 1950.

Fernald, Merrit L., and Kinsey, Alfred, C. *Edible Wild Plants of Eastern North America.* Revised by Reed, C. Rollins, New York: Harper & Row, 1958.

Gerdemann, J.W. "Mycorrhizae." *The Plant Root and its Environment.* Charlottesville: University of Virginia Press, 1971.

Gibbons, Euell. *Stalking the Healthful Herbs.* New York: David McKay Co.,1966.

Gibbons, Euell. *Stalking the Wild Asparagus.* New York: David McKay Co., 1962.

Gray, Asa. *Field, Forest and Garden Botany.* Revised Edition, New York American Book Co., 1887.

Gray, Asa. *How Plants Grow.* New York: Wison, Blakeman, Taylor & Co., 1858.

Gray, Asa. *Lessons in Botany.* Revised Edition. New York: American Book Co., 1887.

Grieve, M. *A Modern Herbal.* Volume 2, republication of 1931 Edition, New York: Dover, 1971.

Grimm, William Carey. *How To Recognize Flowering Wild Plants.* New York: Castle Books, 1968.

Harding, A.R. *Ginseng and Other Medicinal Plants,* republication of 1908 Edition, Boston: Emporium Publications, 1972.

Harris, Ben Charles. *The Compleat Herbal.* Barre, Massachusetts: Barre Publishers, 1972.

Hayden, Ada. "The Ecologic Subterranean Anatomy of Some Plants of a Prairie Province in Central Iowa." *American Journal of Botany.* Volume 6, March 1919.

Hutchens, Alma R. *Indian Herbology of North America.* Ontario: Merco, 1973.

Jacobson, C.A. "Water Hemlock (Cicuta)." Nevada Agricultural Experiment Station, *Technical Bulletin 81.* 1915.

Kingsbury, John M. *Deadly Harvest.* New York: Holt, Rinehart, Winston, 1972.

Kloss, Jethro. *Back to Eden*. New York: Lancer, 1971.

Krochmal, Arnold and Connie. *A Guide to the Medicinal Plants of The United States*. New York: Quandrangle, 1973.

Levy, Juliette de Bairacli. *Herbal Handbook for Everyone*. Newton, Massachusetts: Branford, 1966.

McCleod, Dawn. *Herb Hand Book*. No. Hollywood, California: Wilshire, 1972.

Medsger, Oliver Perry. *Edible Wild Plants*. New York: Collier, 1966.

Meyer, Joseph E. *The Herbalist*. Revised and Enlarged by Clarence Meyer. Indiana: Hammond, 1960.

Peterson, Roger Tory, and McKenny, Margaret. *A Field Guide To Wildflowers*. Boston: Houghton Mifflin, 1968.

Plato, *Phaedo*. Translated F.J. Church. New York: Bobbs Merrill Co. Inc., 1951.

Pond, Barbara. *A Sampler of Wayside Herbs*. Riverside, Connecticut: The Chatham Press, 1974.

Porter, C.L. *Taxonomy of Flowering Plants*. San Francisco: W.H. Freeman, 1959.

Quinn, Vernon (Elizabeth). *Roots: Their Place in Life and Legend*. New York: Frederick Stokes Co., 1938.

Radford, Albert E., and Ahles, Harry, E., and Bell, and C. Ritchie. *Manual of the Vascular Flora of the Carolinas*. Chapel Hill: University of North Carolina Press, 1968.

Rose, Jeanne. *Herbs & Things*. New York: Grosset & Dunlap, 1972.

Schuurman, J.J., and Goedewaagen, M.A. *Method for Examination of Root Systems and Roots*. Wageningen: Center for Agricultural Publishing and Documentation, 1971.

Shosteck, Robert. *Flowers and Plants*. New York: Quadrangle, 1974.

Twentieth Century Alchemist: Legal Highs. San Francisco, California: Level Press, 1973.

Veninga, Louise. *The Ginseng Book*. Santa Cruz, California: Ruka, 1973.

Vogel, Virgil J. *American Indian Medicine*. Norman: University of Oklahoma Press, 1970.

Weaver, J.E. "Investigations on the Root Habits of Plants," *American Journal of Botany*. Volume 12, October 1925.

Weaver, J.E. "The Ecological Relations of Roots." Carnegie Institute of Washington: Pub. No. 286, 1919.

Weiner, Michael A. *Earth, Medicine – Earth Foods*. New York: Collier Books, 1972.

Wilson, Carl L., and Loomis, Walter E. *Botany*. Fourth Edition, New York: Holt, Rinehart and Winston, 1967.

Wilson, Charles Morrow. *Roots, Miracles Below*. New York: Doubleday, 1968.

Wren, R.C. *Potter's New Cyclopaedia of Medicinal Herbs and Preparations*. Re-edited and Enlarged by R.W. Wren. New York: Harper Colophon, 1972.

Yogananda, Paramahansa. *Autobiography of A Yogi*. Los Angeles: Self Realization Fellowship, 1946.

INDEX

Entries in **bold face** indicate main common names. Alternate designations, given in the text as OTHER COMMON NAMES, are shown below in regular roman type.